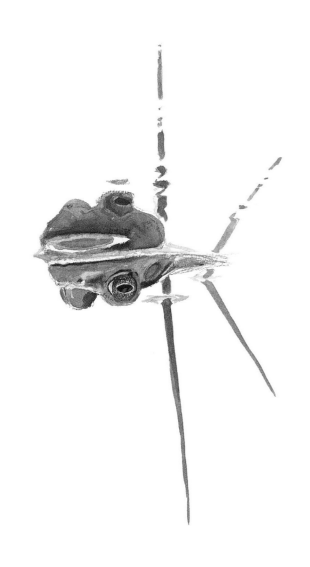

The Nature of Massachusetts

Yellow-rumped warbler and beach plum

MASSACHUSETTS AUDUBON SOCIETY

The Nature of Massachusetts

CHRISTOPHER LEAHY

JOHN HANSON MITCHELL

THOMAS CONUEL

Illustrated by Lars Jonsson

ADDISON-WESLEY PUBLISHING COMPANY, INC.

Reading, Massachusetts   Menlo Park, California   New York

Don Mills, Ontario   Harlow, England   Amsterdam   Bonn

Sydney   Singapore   Tokyo   Madrid   San Juan

Paris   Seoul   Milan   Mexico City   Taipei

Library of Congress Cataloging-in-Publication Data
Leahy, Christopher
  The nature of Massachusetts / Christopher Leahy, John Hanson Mitchell, Thomas Conuel ; illustrated by Lars Jonsson.
      p.   cm.
    At head of title: Massachusetts Audubon Society.
    Includes bibliographical references (p.   ) and index.
    ISBN 0-201-40969-0 (alk. paper)
    1. Biotic communities—Massachusetts 2. Habitat (Ecology)—Massachusetts. 3. Natural history—Massachusetts. 4. Wildlife refuges—Massachusetts. I. Mitchell, John Hanson. II. Conuel, Thomas. III. Title.
  QH105.M4L435 1996
  508.744—dc20   96-15366
                                                        CIP

Jacket design by Jean Seal
Text design by Janis Owens
Set in 10.75-point Minion by Janis Owens
Copyediting by Katherine Scott and Cynthia Benn
Index by Barbara E. Cohen
Production coordination by Pat Jalbert
Prepress by Laser Graphics, Inc.

1 2 3 4 5 6 7 8 9 10-ARK-0099989796
First printing, May 1996

Walking fern

To the founders and members of the Massachusetts Audubon Society,
with gratitude for one hundred years of conservation

Contents

Dear Friends:

Lieutenant Governor Cellucci and I join the citizens of the Commonwealth in celebrating the Massachusetts Audubon Society's 100th anniversary. Since its inception, the Society has helped preserve the extraordinary beauty of the Commonwealth's 8000 square miles, embodying the true spirit of conservation in ensuring that our natural heritage—our forests, waters, and wildlife—is managed wisely and held in trust for succeeding generations.

It is no accident that the Commonwealth has been home to such gifted and influential naturalists as Louis Agassiz and Ralph Waldo Emerson. No one who has watched the dawn strike the high dunes on the outer Cape, pitched a canoe into any of our 28 river basins, or hiked Mount Greylock during the autumn foliage display can deny the raw beauty of Massachusetts.

The Commonwealth's wild landscapes fueled the determination of Harriet Hemenway and Minna B. Hall, founders of the Massachusetts Audubon Society and mothers of the American conservation movement. These courageous women radically changed the way we perceive wildlife. Through their efforts, wild birds and other aspects of nature, once regarded primarily as goods for consumption, came to be viewed as part of a national treasure—a birthright to be restored and preserved for the common good.

The fruits of the first Audubon movement are manifest in Massachusetts today. Thousands of acres of wildlife refuges, state forests, conservation holdings, and local land trusts; a natural heritage program that catalogues and guards our rarest species; the strongest environmental laws in the nation; and, of course, the Massachusetts Audubon Society, still leading the charge after one hundred years, are all part of the legacy of Mrs. Hemenway and Mrs. Hall and their vision of a better society.

This magnificent book captures the Commonwealth's rich natural splendor, commemorating the Massachusetts Audubon Society's Centennial. The Society's continued leadership will help ensure that the treasures featured herein are enjoyed for centuries to come.

Sincerely,

Bm Weld

William F. Weld

For the Love of Nature

INTRODUCTION BY GERARD A. BERTRAND, PH.D., J.D.

President, Massachusetts Audubon Society

THE MASSACHUSETTS AUDUBON SOCIETY, founded one hundred years ago, arose out of a simple passion to right a great wrong. Harriet Hemenway and Minna Hall—our "founding mothers"—looked at their friends' fashionable hats and saw not pretty feathers, but dead birds and the specter of extinction. The outrage that called them to action was not inspired by the waste of a resource or even a sense of personal deprivation—neither of our heroines was an ornithologist—but rather by the intolerable callousness and arrogance of destroying living creatures of great beauty for nothing but greed and vanity. As the conservation movement matured, rational, utilitarian arguments for the protection of the natural world proliferated: Hawks eat our agricultural pests; clean rivers supply our drinking water; trees filter our air…. Such facts are accurate certainly, but insufficient, I think, to have ignited the Conservation Revolution of 1896 and kept it ablaze until bird protection and national parks were the law of the land. No doubt it is partly enlightened self-interest that makes a majority of Americans tell the pollsters that they are environmentalists, but I believe it is mainly the same deep, positive emotion that moved Mrs. Hemenway and Miss Hall: love of Nature.

I have always found it useful in charting the course of the Society or even making routine management decisions, to test my own inclinations against that basic instinct. What would Harriet and Minna have thought? Certainly they would be aghast to discover the zeal and ingenuity we have applied to inventing ever more effective threats to life on the planet. DDT, PCBs, radioactive clouds, supertankers, ozone holes…it makes the bird shooters and milliners of old seem almost innocuous. Regarding the stewardship of the organization they founded, I think they would approve of the fact that on the whole we have stuck to basics: protecting land and wildlife, making nature a part of people's lives, taking bold stands when necessary in pursuit of the mission and ensuring the Society's future.

I hope they would also feel, along with today's staff, board, and membership, great pride in the present scope of our conservation efforts. We now protect 25,000 acres of wildlife sanctuaries from Wellfleet to Pittsfield. This system contains examples of all of the natural communities described in these pages except the open ocean and an astonishing diversity of plants and animals, including hundreds of the Commonwealth's rarest species, some of them of global significance. We are presently adding to our holdings at an average rate of 1,000 acres per year. Our eighteen staffed centers provide more than 5,000 natural history programs a year, and our school programs reach one-fifth of all Massachusetts children in grades K through 12. Our legislative office in Boston is the most credible and therefore the most influential voice for the environment on Beacon Hill, and our regional policy offices in Worcester and on the North Shore help dozens of communities with issues such as water quality, land use, recycling, and habitat protection. We have recently established a Center for Biological Conservation with a staff of professional scientists and naturalists who conduct conservation-based research throughout New England and produce a broad range of scientific and popular publications. Because over half of our nesting birds winter in the Neotropics, we have created partnerships with the people of Belize and Costa Rica, resulting to date in the protection of a quarter million acres of critical habitat. Our regular staff now stands at 175, supplemented by 400 seasonal and part-time employees and over 2,000 invaluable volunteers who give us more than 72,000 hours a year. Our membership has doubled in the last decade to 55,000 households and continues to grow. Far from resting on our laurels, we are presently planning a new international birding and nature center in Newburyport, a new policy office in Barnstable, and our first truly urban sanctuary on the site of the old Boston State Hospital in Mattapan.

It is customary at the hundred-year juncture to prolong this kind of institutional trumpet blowing into an entire volume. Having consulted with the spirits of the founding mothers, however, we have decided to make this book a celebration not of the Society but of that commonwealth of biological diversity on which our mission is focused: The Nature of Massachusetts. The book begins with a brief and graceful history of the first Audubon Society written by John H. Mitchell, editor of our award-winning journal, *Sanctuary*, then treats us to descriptions of the Commonwealth's major natural communities, by Chris Leahy, director of our Center for Biological Conservation, and concludes with vignettes of our sanctuaries by noted natural history writer Thomas Conuel. As you will have noticed long before you turned to this introduction, the whole is spectacularly illustrated by Lars Jonsson, one of the greatest painters of nature alive today.

I believe this is the finest work on the natural history of a state ever published. It is a fitting tribute to the women and men who had the passion, courage, and foresight to found the American conservation movement in Massachusetts a hundred years ago and an expression of gratitude to all who, for the love of nature, have nurtured the Massachusetts Audubon Society.

Lincoln, Massachusetts
February 1996

Acknowledgments

MANY PEOPLE HAVE CONTRIBUTED their expertise on natural communities, species, localities, and other matters to make this book as informative and accurate as possible. We owe special gratitude to those colleagues at the Massachusetts Audubon Society who read parts or all of the natural communities chapter, made suggestions, corrected errors, and filled many gaps and supplied the occasional phrase or paragraph. These include: Jeanne Anderson, Robert Buchsbaum, Betsy Colburn, Peter Dunwiddie, Andrea Jones, Wayne Petersen, and Peter Vickery. Other Massachusetts Audubon staff who contributed to this volume in one way or another are: Kay Beaver, Jerry Bertrand, Jack Clarke, David Clapp, Linda Cocca, Betty Graham, Ann Hecker, Scott Hecker, René Laubach, Simon Perkins, Bob Prescott, and Tom Tyning.

Special thanks are also due to the staff of the Massachusetts Natural Heritage and Endangered Species Program, especially Pat Swain and Henry Woolsey, who made the program's comprehensive database and community profiles readily accessible, answered numerous questions, and supplied summary information available from no other source.

Other colleagues and fellow naturalists to whom we are indebted are: Mark Anderson (The Nature Conservancy), Jim Cardoza (Massachusetts Division of Fisheries and Wildlife), Barry Van Dusen, Tom French (Massachusetts Natural Heritage and Endangered Species Program), Karsten Hartel (Museum of Comparative Zoology), Gwil Jones (Northeastern University), Anne Marie Kittredge (Massachusetts Department of Environmental Management), Jack Lash (Massachusetts Department of Environmental Management), Mark Mello (The Lloyd Center for Environmental Studies), Doug Smith (University of Massachusetts), Tim Simmons (The Nature Conservancy), Leslie Sneddon (The Nature Conservancy), Paul Sommers (Massachusetts National Heritage and Endangered Species Program), Lisa Vernegaard (The Trustees of Reservations), and Bryan Windmiller.

Finally, for giving this book any grace that it now possesses, we are immensely indebted to the editorial expertise of Merloyd Lawrence of Merloyd Lawrence Books and Pat Jalbert of Addison-Wesley Publishing Company.

Since we were not always wise enough to ask or heed the advice of the generous people named above, we will have to take credit for any errors of fact or dubious assertions of opinion.

About the Authors

LARS JONSSON is known throughout the world for his paintings of wildlife. Like the work of the great masters, J.J. Audubon, Bruno Liljefors, or Louis Agassiz Fuertes, his art transcends the genre of "wildlife art" and is exhibited and praised in museums throughout Europe and the U.S. His many beautiful books include *Birds of Europe*, now translated into seven languages. Lars Jonsson lives on Gotland, an island off the coast of Sweden.

JOHN HANSON MITCHELL is the author of several highly praised works of natural and human history including *Walking towards Walden: A Pilgrimage in Search of Place*, and three books that focus on a single square mile of land where he lives in Eastern Massachusetts: *Ceremonial Time, Living at the End of Time*, and *A Field Guide to Your Own Backyard*. Editor of *Sanctuary* magazine, he is the winner of the 1994 John Burroughs Essay Award.

THOMAS CONUEL is a natural history writer with a special interest in rivers, land use, and water issues. He is author of *Quabbin: The Accidental Wilderness*, the unique story of the wildlands created when the water authority flooded a vast area of Central Massachusetts. He has worked as a reporter, technical writer, and newspaper editor and lives in Petersham, Massachusetts.

CHRISTOPHER LEAHY is Director of Massachusetts Audubon Society's Center for Biological Conservation, overseeing ecological management, field ornithology, and biological atlas programs. A recognized authority on birds and insects, he is the author of *The Birdwatcher's Companion* and *The First Guide to Insects*. When not leading expeditions to outer Mongolia and other far corners of the world, he lives in Gloucester, Massachusetts.

Lady's-slipper

Bald eagle over Quabbin Reservoir

The Nature of Massachusetts

The Founding Mothers

Snowy egret

There is a breed of New Englander, as yet not extinct in our time, which commonly frequents well-known bird-ing spots in the region. The species evolved in Boston sometime in the late seventeenth century and has somehow managed to thrive in spite of changing economies, rampant, unpredictable technologies, an ever-changing political climate, and violent upheavals in the socioeconomic landscape. I spotted two members of this group not long ago standing beside their car near one of the pull-outs by the marshes at the Parker River National Wildlife Refuge. The car was a modest, nonde-script, older model with a camp blanket in the backseat, much upholstered with the hair of a white dog, and assorted stickers of environmental organizations plastered on the back window. The male was in his midsixties, with close-cropped gray hair and piercing blue eyes, and wore a tat-tersall shirt, rumpled gray flannels, and worn boat shoes. The female, presumably his mate, wore a wool skirt, a blue turtleneck, a short wool jacket, and tennis sneakers. They both carried antique, well-used binoculars, dating, I would judge, from the early 1960s, when they must have been in their birding prime.

While we talked about weather fronts and birds, they eyed the upper air and the marshes. Just as we were about to part company, the woman said something about tree swallows and began making squeaking noises at the sky. Within a second or two, a group of four tree swallows darted down and circled around her, followed by three more. Her husband joined in, and soon the sky was filled with diving birds. They appeared as if from nowhere, a great locustlike swarm, cheeping and twittering, and executing sharp turns above the dunes and swales.

"Quite a storm," the woman said.

This whole affair was completed in a busi-nesslike manner, without ceremony or fuss, and, when the flock swirled out across the marshes and disappeared, the couple began searching the salt pannes for shorebirds, with no-nonsense effi-ciency. It was as if they were somehow responsible for the well-being of this place and its birds—which, in a manner of speaking, I suppose they were.

Women of this group were once dreaded by highway builders and developers. They were known troublemakers, dogged fighters, ever po-lite, sometimes halting in speech, given perhaps to digression, but blessed with such an unstoppable ability to stick with a cause that they could, with the help of well-placed allies, bring down giants of industry. This breed was once, and to some ex-tent still is, a known entity in Washington—the scourge of certain senators from the West, the bane of industry lobbyists, pesticide manufactur-ers, timber company officials, water managers, special-interest groups, and anyone else with a grand scheme for the makeover of the natural en-vironment.

During the 1950s, a woman of this ilk named Olga Owens Huckins lived on a small property in Duxbury, Massachusetts. She had turned her land into a private bird sanctuary, well stocked with feeders and birdbaths. But in 1958, as a result of a state-sponsored aerial mosquito spraying of the South Shore, many of the birds, including those that had nested on the property year after year, were killed outright. Huckins had solid documentation of the event and wrote an impassioned letter to the *Boston Herald,* detailing what she viewed as an inhuman, undemocratic, and even unconstitutional act.

Because she foresaw the possibility of the increased use of aerial spraying, Huckins sent a copy of the letter to a woman she knew who worked with the U.S. Fish and Wildlife Service, in which she inquired about persons in Washington who might be able to help. Huckins's contact, who had been tracking pesticide-related incidents of this sort for years, made inquiries, but, in the end, could not locate anyone who would be able to assist her. The woman decided to do something about it herself. She began assembling her notes and data, spent the next two years doing further research, and finally wrote a book. The woman was Rachel Carson; the book was *Silent Spring.*

A point that has generally been overlooked by historians and analysts of political currents in this country is that historically, environmental activism has been the work primarily of women. Theory, philosophy, and writing have been the handiwork of men such as Henry Thoreau and John Muir, but action—that is, letter-writing campaigns, organization, boycotts, demonstrations, the willingness to lie down in front of offending bulldozers, and the like—has been the business of women. Marion Stoddard, for example, the woman who was primarily responsible for the salvation of the Nashua River, had a blunt, direct style of confrontation. Once she brought a bucket of vile, polluted, river water into the governor's office and placed it on his very desk. On another occasion, she and her children wrote a short, simple missive to the governor—in large letters: "We would like to remind you," it said, "that the Nashua River *stinks.*"

One of the seminal events in the history of environmental activism in this country took place in a parlor in Brookline in 1896. On a January afternoon that year, one of the scions of Boston society, Mrs. Harriet Lawrence Hemenway, happened to read an article that described in graphic detail the aftereffects of a plume hunter's rampage—dead, skinned birds everywhere on the ground, clouds of flies, stench, starving young still alive in the nests—that sort of thing. The slaughter was in the service of high fashion, which dictated in those times that ladies' hats be ornamented with feathers and plumes, the more the better.

Harriet Hemenway was properly disturbed by the account, and inasmuch as she was a Boston Brahmin and not just any lady of social rank, she determined to do something about it. She carried the article across over to the house of another social luminary, her cousin Minna B. Hall. There, over tea, they began to plot a strategy to put a halt to the cruel slaughter of birds for their feathers. Never mind that the plume trade was a multinational affair involving millions of dollars and some of the captains of nineteenth-century finance; the two women meant to put an end to the nasty business.

Harriet Hemenway, it was often said, had a mind of her own. She once entertained a black man as houseguest when he could not find lodging elsewhere in Boston (he happened to be Booker T. Washington, but that is beside the point). She would fire off public denunciations of other Brahmins if she felt they needed correction, and, when she sat for John Singer Sargent for her portrait, she let the world know she was pregnant by holding a water lily to her breast—symbolic language proclaiming her condition, and a rare, even shocking, public announcement for the period. She was independent, a bit of an iconoclast, an activist, boundlessly energetic, gregarious, overly fond of chocolate and tea, and, furthermore, she lived for a very long time. Not a few people around today remember her.

Boston tended to produce such women. Unlike members of well-heeled families of other cities, when Bostonians came into money, instead of constructing grand estates in Newport or the Hudson River Valley, they had a predilection for putting their riches into educational institutions, schools for the blind, social service institutions in mill cities such as Lowell, and other good works. It was Boston money that built some of the first museums and libraries in this country. It was Boston money that backed the abolitionist movement, and, when the war finally came, it was a Boston family that put one of its favored sons, Robert Shaw, at the head of a company made up entirely of African Americans. Harriet Hemenway, née Harriet Lawrence, was a product of this tradition. She came from a family that had made money from the textile mills. Her father was a devoted abolitionist and a great supporter of education, and, quite naturally, when she came to marry, it was only right that she do so within the Brahmin clan. She became a Hemenway, joining another illustrious, rich, and active Boston family.

The Brahmins had a deep moral streak, part of which was no doubt inherited from their Puritan forebears. But that is not to say they were without sin. They had made their money in the satanic mills, in the China trade, or even in the slave trade, and, by 1896, many of the female family members, Harriet Hemenway and Minna Hall included, would commonly wear upon their hats the elaborate nuptial plumes of murdered birds. The practice was so common in the United States that there were fewer than five thousand egrets nesting. Terns had been entirely extirpated from the southernmost New England states, partly because of the rage for plumed hats, and, by 1896, it was estimated that some five million American birds of about fifty species were being killed annually for fashion. But unprincipled acts such as the wanton slaughter of innocent birds for so shallow a matter as fashion would not long endure once Harriet Hemenway was on the case.

She and Minna Hall took down from a shelf *The Boston Blue Book,* wherein lay inscribed the names and addresses of the members of Boston society. Hemenway and Hall went through the list and ticked off the names of those ladies who were likely to wear feathers on their hats. Having done that, they planned a series of tea parties. Women in feathered hats were invited, and, when they came, over petits fours and Lapsang souchong, they were encouraged, petitioned, and otherwise induced to forswear forever the wearing of plumes. After innumerable teas and bouts of friendly persuasion, Harriet and Minna had established a group of some nine hundred women who vowed "to work to discourage the buying or wearing of feathers and to otherwise further the protection of native birds." Hunters, milliners, and certain members of Congress may have found the little bird club preposterous. After all, the feathers were plucked from long-necked things that lived in swamps and ate tadpoles, as one senator would later phrase it, whereas their plumes decorated the hats of beautiful ladies. The Lord made birds for bonnets, it was argued.

But opponents of any regulation on the trade underestimated their opposition. The Boston club was made up of women from the families of the Adamses and the Abbots, the Saltonstalls and the

Snowy egret

Cabots, the Lowells, the Lawrences, the Hemenways, and the Wigglesworths. These were the same families that brought down the British empire in the Americas. This was the same group that forced Lincoln to issue the Emancipation Proclamation, and it was these families that were about to create the American tradition of environmental activism. Within a matter of decades, the little bird club had spawned what would be the most influential conservation movement in America up to that time.

Notorious, independent Boston women notwithstanding, these were not the freest of times for society women, and Hemenway and Hall were wise enough to know that if their group were to have any credibility it would need the support of men, and most importantly, would need a man as its president, even if he would be a mere figurehead. The women organized a meeting with the Boston scientific establishment, outlined their program, and got the men to agree to join the group, which would be called, they decided, the Massachusetts Audubon Society, in honor of the great bird painter John James Audubon.

From the start, the organization had the backing of some outstanding names in American ornithology: Edward Howe Forbush, George Mackay, and the naturalists Charles S. Minot and Outram Bangs. Minot was associated at Harvard with the foremost biologist in America, Louis Agassiz. The women made Minot chairman of the board. Then they chose as their president one of the cofounders of the Nuttall Ornithological Club and the American Ornithologists' Union, the Cambridge bird man, William Brewster. It was a skilled political choice. With Brewster as head of the organization, the Society immediately garnered national recognition—which is what the women wanted. This was, after all, a national issue; birds were killed and hats were worn in most of the forty-six states.

By the third meeting of the young organization, the board resolved to use every effort it could to establish similar societies in other states. By 1897 Pennsylvania, New York, Maine, Colorado, and the District of Columbia formed groups. Massachusetts began producing leaflets and helped distribute the legislative models prepared by the American Ornithologists' Union to other societies. By 1900 a conference of state Audubon societies was held in Cambridge, and, the following year, Massachusetts organized another conference in New York. By 1905, with the prodding and money of the Brahmin women, a national association of Audubon societies was established, a group that eventually became the National Audubon Society. All this was to accomplish a single purpose: to do something about the continued slaughter of plume birds. By 1897 Massachusetts had passed a bill outlawing trade in wild-bird feathers, and in 1898 Massachusetts Senator George Hoar attempted to introduce a bill to the U.S. Congress to prohibit both the sale and shipment of plumes within the United States and their import or export to other nations. The bill failed, but sentiment for the cause was running strong by this time, and when Congressman John Lacey of Iowa proposed a bill in 1900 to prohibit the interstate shipment of animals killed in violation of local state laws, it passed. The Lacey Act, coupled with strong state bird-protection laws and the use of agents to enforce them, slowly began to weaken the trade, or at least make it more difficult. In addition, the fact that a friend of Minot's family and a former member of Brewster's Nuttall Ornithological Club, Teddy Roosevelt, became president of the United States in 1901 certainly did not harm the cause.

Like many successful campaigns, this one was fought on two fronts. Laws were passed, and, just as important, social pressures were applied. In 1909, when the first lady, Mrs. William Howard Taft, had the audacity to appear at the presidential inauguration with feathers in her hat, Minna Hall promptly wrote her a personal letter of remonstrance. Brahmin women had never been considered paragons of fashion: they shopped at R. H. Stearn's and preferred pearl chokers, low-heeled shoes, and long-sleeved nightgowns. By 1920, no lady with any sensibility would be seen on the streets of Boston wearing feathers, at least not

without being glared at by one of her sisters. It was not unlike the current movement against wearing furs.

From time to time the legislative and the social fronts converged. In one fight in New York state, a deluge of letters and petitions from women's clubs convinced legislators to pass laws restricting bird hunting. A bill was passed in 1913 to protect migratory birds, and by 1916 the Migratory Bird Treaty with Great Britain further reinforced the law. By the 1920s the issue was dead. The trade had been made illegal, and, although Harriet Hemenway would still have to glare at an occasional offender on the streets of Boston as late as the 1940s, feathers were going out of fashion anyway.

But the fight was not over. There was the matter of that other phrase in the founding charter, to "otherwise further the protection of our native birds," and Harriet Hemenway, who was sixty-two in 1920, still had another forty years to go. Proper Boston women, it used to be said, *liked* getting old. They could wear their hair in the Queen Mother style with impunity, ignore fashion altogether, and say what they wanted. The Boston abolitionist Julia Ward Howe, who wrote "The Battle Hymn of the Republic" and lived to be ninety-one, confided to her diary at the age of eighty-seven that she hoped the coming year would bring her *useful work*. Aging was like a cup of tea, she believed. The sugar was at the bottom.

THE GREAT ECOLOGICAL THEORIST Henry Thoreau believed that every town should have a plot of wild land preserved for instruction and recreation, where not a stick of firewood should be cut and where we could sense something of the older, wilder order. Thoreau was always out of step, but he was often ahead of the parade. When it appeared that the battle of the plumes was going to be won, members of the Massachusetts Audubon Society began thinking of other ways to protect native birds. Quite naturally, the progenitors of conservation, Thoreau and Emerson, came to mind, and when, in 1916, a prominent society man, George W. Field, offered the use of his estate in Sharon as a bird sanctuary, the Society felt it could be a place of instruction as well as sanctuary and decided to turn the land into an educational center. The property (which was in a different location from the current Massachusetts Audubon Moose Hill Wildlife Sanctuary in Sharon) consisted of some 225 acres of diversified fields and forests, complete with running brooks and a pond. In 1918 the Society appointed a warden, Harry Higbee, to maintain the property and demonstrate to visitors how birds may best be attracted to any farm or estate. Higbee lived on the property in an ancient farmhouse, which he opened to the general public. In separate rooms in the house, Higbee displayed his collection of birds, minerals, flowers, and insects, as well as the ever-present and well-distributed Audubon literature. By 1920 the little sanctuary had become a sort of mecca for those interested in natural history. The site had over twenty-six hundred visitors that year, from some twenty-three states and not a few foreign countries, including England and Cuba; there was even a delegation of Japanese studying American institutions—of which, by 1920, the Massachusetts Audubon Society was one. In fact, membership in the Society—along with the Boston Symphony, the Massachusetts Horticultural Society, the Boston Athenæum, and the Chilton Club—had become *de rigueur* for the Brahmins.

In 1922, determined to build on the success of its first sanctuary, the Society purchased more land in Sharon, a forty-three-acre site that became the core of the present two-thousand-acre Moose Hill Wildlife Sanctuary. The concept of land as an administrative base, a staging area for education, and a sanctuary to protect wildlife was established. It was a new twist. There were, by the 1900s, thousands of acres of preserved open spaces in New England; there were political organizations fighting the proto-environmental battles of the era; and there were even a few early outdoor educational programs. But until that time, no organization had managed to combine all three under one banner.

This idea of saving land to protect species was hardly new. Marco Polo reported that the great thirteenth-century Mongol ruler Kublai Khan had set aside a vast holding of land near the northern city of Changa-nor for the sole purpose of preserving and even producing wildlife. Never mind that the primary reason for this was to provide the grand Khan and his noblemen with successful hunting—the sanctuary, although it wasn't called that, was a successful and innovative idea. During the same period, Europeans were also setting land apart from normal human activities in the form of the royal deer park. In America, it was not until the midnineteenth century that the idea of setting aside grand tracts of land for conservation purposes was established.

Here in America, we at least had an excuse for not saving land—there was so much of it. We had to destroy it before we could think about preserving it. But even as early as 1820, it was clear that the burgeoning industrialization in the East would in time overwhelm the entire continent. Thoreau and Emerson recognized the threat, as did painters such as George Catlin and Thomas Cole. Catlin, the great painter of Indians, foresaw the fate of the nation and its native peoples and recommended setting aside a vast "park," as he called it, where the natives could continue to live undisturbed.

The Englishman Thomas Cole took a sketching trip up the Hudson River Valley in 1820 and was so impressed with the primal beauty of the river and nearby Catskills that he settled in the area. Soon other artists followed him to the region, and in time a recognized school of landscape painters was formed, the Hudson River School, the first such group in America. By the time of the Civil War, painters out of this tradition had discovered the vast pictorial landscapes of the American West. By 1872, when Congress voted to establish a "national" park at Yellowstone, the public was ready for the move.

New England, as usual, was ahead of the pack. In 1831, Mount Auburn Cemetery, America's first "garden" cemetery—essentially a landscaped open space—was established in Cambridge. By 1853, the residents of Stockbridge, Massachusetts, founded a group known as the Laurel Hill Association, which set aside a wooded hill in the center of the village as a wildwood park. By the next decade, under the direction of park planners Frederick Law Olmsted and Boston's native son Charles Eliot, the concept of an urban common as wild park—a place that at least imitated a natural landscape—was coming into vogue. Olmsted and Eliot's Emerald Necklace, originally designed to surround the city, eventually preserved some 2,200 acres of green space. What began in Boston with Olmsted's firm soon spread throughout the Northeast, as city after city began to set aside open spaces and establish landscaped natural parks. It is certainly one of the great ironies of our time that the city with one of the highest percentages of preserved open space in the United States is New

York, thanks mainly to Olmsted's innovative Central Park. The place is now a popular birding site.

By 1876, the Appalachian Mountain Club was organized in Boston to support the preservation of open space throughout New England and expand opportunities for outdoor recreation, which in this period was becoming the rage of Boston society. By 1891, a group of Brahmins who came from the same patrician families that would make up the board of the Massachusetts Audubon Society a few years later voted to start an organization known as The Trustees of Reservations, whose purpose was to acquire and maintain beautiful or historic sites throughout the Commonwealth. It was the first organization in the world devoted solely to the preservation of open space for public purposes, and it became the model, a few years later, for Britain's National Trust, which has been such a major force for the preservation of the historic and natural landscapes of Britain.

The effort to save land has borne fruit. The Trustees of Reservations now hold 19,500 acres. The New England Forestry Foundation has 15,000 acres throughout New England. The Commonwealth of Massachusetts holds some 500,000 acres, and towns maintain almost another 100,000 acres of open space. The 55,000-member Massachusetts Audubon Society, which is now the largest conservation group in New England and one of the largest in the nation, maintains 24,000 acres, including eighteen staffed sanctuaries.

This idea of saving land to protect birds had taken hold by the 1920s, and over the next four decades some of the major properties held by the Society were donated: Moose Hill in Sharon, Arcadia in the Connecticut River Valley, Pleasant Valley in the Berkshires, Ashumet and Wellfleet on Cape Cod, Wachusett Meadow in central Massachusetts,

Roseate terns, courtship display

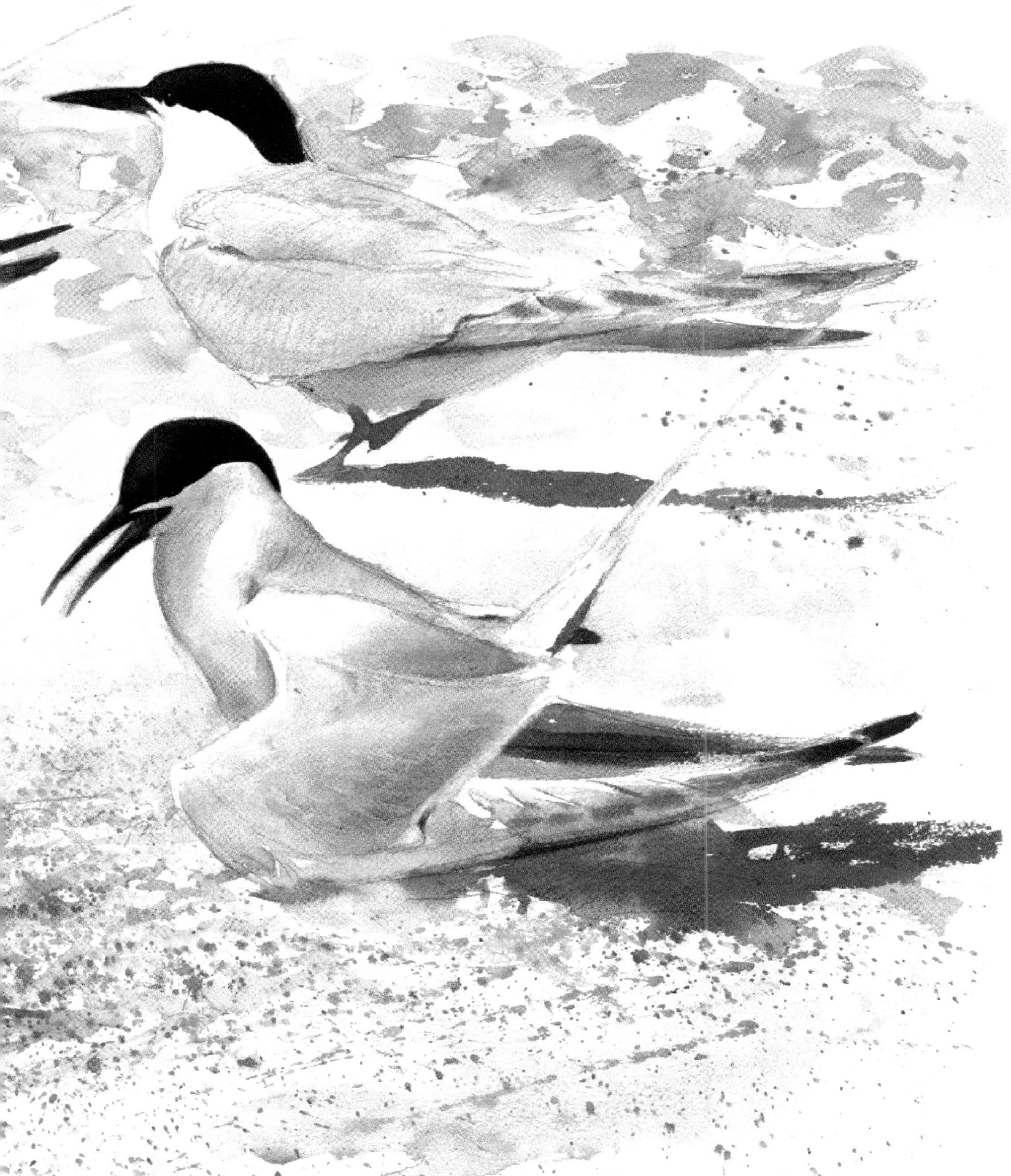

and Drumlin Farm in Lincoln, just west of Boston. But in spite of a period of steady acquisition, these were quiet years. Other organizations in the country, including a few seedlings of the original Boston group, were beginning to flourish, whereas in Massachusetts the members were more interested in—one might even say obsessed with—birds. All quite natural. From the beginning some of the greatest names of American ornithology had been associated with Massachusetts Audubon—William Brewster, Charles W. Townsend, Edward Howe Forbush, the great bird artist Louis Agassiz Fuertes, Arthur Cleveland Bent, and also the dean of all birdwatchers, Ludlow Griscom, who championed sight recognition as a means of identifying birds in the field and, up until the publication of Roger Tory Peterson's field guide in 1934, did more to foster popular birdwatching in this country than any other individual.

There is, nevertheless, a certain redundancy in the biographies of those young, energetic bird men who were now associated with the Massachusetts Audubon Society; you could almost write one for them all. Born: eastern Massachusetts. Schools: Milton Academy, Harvard. Place of residence: North Shore. Interest: ornithology.

The same might be said of the programs offered by the organization in these quiescent years. Witness this announcement from the 1921 Massachusetts Audubon Society Bulletin:

MARCH 29, 1921. *The Audubon Bird Lectures will be given again this year in Symphony Hall on Saturday afternoons at 2 o'clock. The Audubon Society's motion picture film "The Birds of Killingworth" will be shown at one of these lectures. The course will be a treat for bird lovers, exceeding anything which the society has yet put before the public. As always, Mr. Edward Avis* [no pun intended, this was apparently his real name], *well known for his wonderful whistling reproductions of bird music, will appear at two of the lectures.*

Notice the use of the term "Audubon Society." In the mind of Boston, there was only one Audubon Society, or at least only one *real* one. In fact, there were dozens all over the country, including the national one, and they were all at work educating schoolchildren about the value of local birds. It was out of this effort that the next generation of conservationists would be fostered.

By 1931, HARRIET HEMENWAY had moved to the family summer house in Canton, southwest of Boston. Her husband, Augustus, was dead. She was seventy-four years old and had a burgeoning crowd of grandchildren. She had not changed her style very much since the 1890s; in fact, it could be said that in some ways she was just coming into style. She dressed in black. She wore a Queen Mary hat (sans plumes, needless to say), and she went about in sensible shoes, which fashionable ladies of Philadelphia referred to as "ground grippers." For outings, she was not embarrassed to don white tennis shoes, an item that in later years would become the very symbol of female conservationists and an object of derision by *New Yorker* cartoonists and threatened developers. Every Friday she took the train to Boston to attend the Chilton Club lecture and luncheon, followed by the Symphony. There she would meet her fellow "low-heelers"—as proper Boston women were called. They were all of them aging now, but, true to Boston form, they were coming into their prime.

On Sundays at one o'clock, the *grande dame*

presided over dinner at the Canton house. These affairs had an English silver, family silver, joints of beef with horseradish sauce, green peas and mashed potatoes. And after dinner—chocolate. Harriet Hemenway adored above all a bit of post-prandial chocolate. Later in life she took to feeding upon strawberries, and when family members reminded her that her doctors had told her that strawberries were bad for her rheumatism, she reminded her family that those doctors were now long dead, whereas *she* was alive and well.

All the Boston doyennes professed a love of nature, at least publicly. (Privately, many of them preferred the shadowed corners of their Victorian sitting rooms, a cheery fire, a cup of tea, and a good book.) But Harriet Hemenway meant it. Survivors of her Sunday outings remember being dragged along on bird walks or stargazing parties—even in the depth of winter. She was always out-of-doors, and she walked everywhere, sometimes to the dread of those family members who began to worry about her health. She was intractable. You could not tell Harriet Hemenway that she was too old to safely walk the grounds. Furthermore, she never gave up on her causes. She started a fund to buy more land. She instructed the New England Forestry Foundation—which by 1953 would be under the direction of her grandson John—to purchase 495 acres in memory of her late husband. She fought for the rights of working women, and she dutifully attended the Saturday-afternoon lecture series of the Audubon Society at Symphony Hall (although it is possible that after all the *Sturm und Drang* of the early years, she was a bit bored—one wonders what she thought of Mr. Avis and his bird whistling).

In 1957, Harriet Hemenway turned one hundred. The Society presented her with a scroll and pointed out that few people leave such significant memorials to their personal foresight, which anywhere else but Boston might have been true.

Six years earlier, in 1951, her co-conspirator and cousin, Minna B. Hall, had died at the age of ninety-two. She had remained active to the end, still serving on the board of directors of the Society, still watching birds, still taking excursions into the woods around Hall's Pond on Beacon Street in Longwood, where she had been born and raised. The pond used to be known as Swallow Pond, and Minna herself was one of the last members of a club of Brahmins, founded by one of the Lowells, known as the Society of Those Still Living in the House They Were Born In. Only a few years before Minna died, Harriet Hemenway confided to one of her fellow Audubon members that Minna was overdoing it. She was too busy, Harriet said. Minna countered. She cornered someone at Symphony one Friday afternoon and suggested that Harriet was gadding about too much and looked tired. "She doesn't know when to stop," Minna said.

These, mind you, were women in their nineties. A few years after Minna's death, Harriet Hemenway broke her hip and couldn't get about as much as she used to. She was still clear-headed, but the racehorse was penned. Then, in 1959, her Society hired as its executive vice-president Allen Morgan, a young insurance executive with an interest in birds and wildlife who was already a member of the Massachusetts Audubon Society board of directors. In this period, on summer nights in the suburbs, strange vehicles with sprayers mounted on flatbed trucks would pass to and fro along quiet town streets, drenching the local elm trees with

Snowy egret

DDT. The practice of aerial spraying for mosquitoes was coming into vogue. Levittown, the first large tract-housing development, had been built, and the design was spreading to Massachusetts. The first shopping mall in the country, on Route 9 in Natick, had opened for business. Coastal and inland wetlands were being filled at such a rate that nearly half of the original wetlands were now drylands. Songbirds were dying horrible deaths on front lawns. And to the dismay of aging traditionalists such as Harriet Hemenway, parking lots were overwhelming green space and historic buildings throughout New England. It was not the best of times.

By 1960, the term "Brahmin," originally coined by Oliver Wendell Holmes to describe those ascetic, intellectual, and cultured aristocrats of Boston, had been in use for about one hundred years. The Boston Brahmins themselves had fled the city and dispersed north and south to live quiet lives on their trust funds. R. H. Stearn's closed its doors forever. Bostonians actually began to admit to having money; some of them even displayed the fact by driving expensive cars, building monumental houses in the suburbs, and wearing jewelry and fashionable clothes. Things were changing, and out in Canton, as if to put it to an end to an era, Harriet Hemenway quietly succumbed. She was one hundred and three years old.

THIS ALLEN MORGAN was a sharp birder and he knew a thing or two about flowers and trees, but he was not, or at least claimed he was not, a real naturalist, as were some of the old-style heads of Massachusetts Audubon. Morgan had been a marine during World War II, and in spite of a mild obsession with bird life, he had become more concerned with the human condition as a result of his experiences during the war. It was he who came up with the little advertising slogan that was once used by the Massachusetts Audubon Society, "We're not just for the birds...."

This was entirely in keeping with the mood of the nation. More and more people were recognizing the relevance of environmental issues to their lives. By the mid-1960s, the Sierra Club, originally founded as a mountaineering group, was actively campaigning on environmental issues. Massachusetts Audubon's large and energetic offspring, the National Audubon Society, was taking a stand against pesticide use; even hunting and fishing groups such as the Isaak Walton League and the newly formed Ducks Unlimited were working to conserve land; and new, active grassroots organizations were appearing across the country to protect open space.

Morgan saw the need to broaden the scope of the Society's activities to meet new challenges. He began to garner friends and allies in high places. He was outspoken and well informed, a gadfly in the style of Harriet Hemenway, and he was not averse to moving the organization into the nasty snake pit of the political arena, which was of course the original intent of the Audubon women. In 1956, Mrs. Louise Ayer Hatheway had donated her entire property and estate, Drumlin Farm in Lincoln, to the Society. When Morgan took over, the organization headquarters were located in a small office on Newbury Street. Under his direction the operation was moved to Lincoln, and the staff began using the farm for educational purposes.

One director at Drumlin Farm, Dr. William H. Drury, was a Harvard biologist and ornithologist with an interest in original scientific research, and he suggested to Morgan that the Society use the sanctuaries for data gathering and controlled studies. After some negotiation, Morgan and Drury reorganized the Society and invited local and international ornithologists to use Drumlin Farm for their studies. The Society had now added a third direction to conservation and education—research.

Initially the research was purely ornithological: nesting behavior of white-breasted nuthatches and chickadees, breeding strategies of killdeer, and the like. But by the early 1960s the scientists began to focus on pesticide use. Morgan was a friend of Olga Huckins's in Duxbury; in fact, he had met Rachel Carson on the property one afternoon when she came up from Washington to look at the land where the story had begun. After the publication of *Silent Spring* in 1962, the Society began to do its own studies on pesticides, under the direction of William Drury. The great furor at this time was the fact that all the evidence was circumstantial. The spray trucks would pass, and the next day robins would be seen quivering and dying on people's front lawns. The lawyerly pesticide companies were quick to point out that, in a court of law, this would be inadmissible evidence—merely coincidence. Furthermore, no one in the state labs could find any residue of pesticides in the bird corpses that were brought in for analysis.

"That," said Morgan, "is because they were using the wrong technique."

In these years there was a new technology on the horizon for detecting pesticide traces: the gas chromatograph. Within the Society, Morgan raised funds to purchase two gas chromatographs, which the Society donated to the Commonwealth of Massachusetts. By now Morgan had hired an assistant, James Baird, to help work on the new envi-

ronmental issues. Baird started on the problem of controlling pesticides, which, since the publication of *Silent Spring*, had finally become recognized as a national environmental problem. Under the prodding of Baird, state labs carried out the tests, and there, as everyone suspected, stored in the birds' fat, was the evidence: the residue of sprays.

Morgan skimmed by on intuition. He had an excellent sense of timing for issues and solutions, and he was not afraid of innovative theories that had yet to be accepted by the scientific establishment. In the early 1960s he read an obscure article about a salt-marsh study by an ecologist named Eugene Odum. The study identified the relationship between energy and food production in salt marshes and offshore fisheries, and Morgan believed the study made a strong case for the importance of halting the draining, filling, and polluting of coastal wetlands. He called his friends and allies in the legislature, got the Society to publicize the story, buttonholed the activists, and gave speeches on the subject to women's groups. He began working to develop legislation with Charles H. W. Foster of the Massachusetts Department of Natural Resources, who had also been studying the issue. Foster had recently instituted a governor's commission on wetlands preservation, and, working together, the two of them managed to get a bill to control draining and development of coastal wetlands through the Massachusetts legislature. It was the first salt-marsh-protection bill in the nation.

From the studies instituted by Foster, Morgan inferred that the same dynamics that made salt marshes a factor in storm-damage control must also be at work in freshwater wetlands. Morgan had lived along the Sudbury River for years and had watched the normal flooding cycles, and it struck him that natural wetlands act like sponges, absorbing waters during heavy rains and slowly releasing the excess over time. So he began pushing for a freshwater protection bill. Once again he sent out his point man, James Baird, who with the assistance of a Department of Natural Resources legal adviser, Robert Yasi, and the legislator Francis Hatch, wrote the legislation that eventually was passed as the Hatch Act, the first law in the country aimed at protecting freshwater wetlands. Some years later these two laws were combined to create the current Massachusetts Wetlands Protection Act, passed in 1979, which was used as a model for federal wetlands protection laws.

Around 1971, Morgan got the idea that the question of energy production was one of the most fundamental and insidious aspects of the multitudinous human effects on the environment. The oil spills that were befouling the nation's beaches were only part of the problem, he believed. On his recommendation, the Society's board hired MIT physicist James MacKenzie as energy specialist to work on energy conservation issues. In 1972 this was something of an outrageous step for an organization that was still, in the minds of the general public, associated strictly with birds. But in 1975, after the oil embargo, the general public began to understand just how pivotal the question of energy consumption and conservation really was. By 1978, when Morgan had a solar heating system installed as a model project at Drumlin Farm and began promoting conservation and solar power as an alternative to fossil fuels, the public understood the ecological connection.

One of the things that was recognized by Harriet Hemenway and Minna Hall was that the issue facing them in 1896 was not local; it was a na-

tional, even international, problem. It was for this reason that the women began providing money to start up other state Audubon societies, worked at the federal level, and through the years devoted so much time and energy to the education of children. Since that time, the awareness of global interconnectedness has only deepened. This realization prompted the board in 1980 to hire as a successor to Allen Morgan a thirty-six-year-old scientist from Washington named Jerry Bertrand, who already had broad international environmental and legal experience. Late-twentieth-century geopolitics, the existence of multinational corporations with no particular allegiance to any nation, and the bioregional nature of environmental issues made the choice logical.

The fact that the Society has had only six presidents in one hundred years says something about the singularity of purpose of the organization. But it is curious that the board seems to pick people with boundless energy to head the Society.

Like that of the earliest presidents and board activists at the fledgling organization, Bertrand's background was academic. Unlike the others, he actually had *two* advanced degrees: a doctorate in biological oceanography from Oregon State University and a J.D. in environmental law from the University of Wisconsin. In addition, he had trained as a civil engineer. He had served as chief of international affairs for the U.S. Fish and Wildlife Service in the Department of the Interior and had dealt with worldwide wildlife problems in Asia and South America. But he had learned his lessons well in environmental geopolitics and knew that all global issues are local issues. So he started to buy land in the state of Massachusetts. In the first eighty-five years of its existence, the

Massachusetts Audubon Society had acquired or been given approximately 11,000 acres of land throughout the state. In the first ten years of Bertrand's reign, he added over half that amount again, about 7,000 acres, and in the next five years another 5,000. Currently the Society has 24,000 acres. Jerry Bertrand's grand design is to have one Massachusetts Audubon sanctuary within twenty minutes' driving time of every urban center in the state and, once the land is secure, to establish nature centers and develop a nature education program that will reach every child in the state. You can always raise money, he says, but you have to save the land while it's still there.

As any ecologist will tell you, however, the creation of relatively small tracts of land (the average size of a Massachusetts Audubon sanctuary is 1,100 acres) is insufficient if you want to save birds and mammals and protect the entire fabric of an ecosystem. By 1980, partly as a result of the research of island biogeographers, it had become clear that with the incredible spurt in development in the Northeast, haphazard, town-by-town zoning, and increasing human population density, in time the preserved open spaces would become totally surrounded by development and begin to function as islands. Islands are notoriously precarious places for local wildlife; furthermore, many of the species, such as wood thrushes, are not suited to small isolated existences. Cut off from traditional migratory routes, access to water, extensive stretches of deep woodland, or old trees, many species of birds, mammals, and even insects would be facing extinction.

Bertrand took seriously the Society's mandate to do everything possible to "further the protection of native birds." He extended his view beyond the boundaries of the Society's sanctuaries and sought larger sections of open space that would interconnect with other large preserved tracts of land, even if the latter didn't belong to the Society. In this way, he aimed to link the many little islands that have been saved in New England by various organizations into a great network of corridors where wild species could travel freely.

Bertrand extended this approach to research, and, in 1993, organized the Center for Biological Conservation, which studies ecological issues on a regional basis. The Society's biological research on local wildlife populations, such as grassland birds, no longer ceases at state borders. As local, state, and national conservation activists have learned in the past thirty years, nature ignores political boundaries.

For a large percentage of Massachusetts and New England birds, even this enlightened regional approach might not be enough. Many birds that inhabit the backyard woods and fields of Massachusetts are in fact mere visitors. Some species, such as the shorebirds and terns, which played so great a role in the foundation of the Society, spend only a few months out of the year here during the summer breeding season. It was clear to Bertrand that if you are going to follow the mandate of the organization and do everything you can to protect "native" birds (whatever that means), you have to go outside of the country to those areas where "our" birds spend most of their lives—which is to say the Neotropics.

During this same period of the eighties, frightening statistics on the decline of natural habitat in the tropics and subtropics began to emerge. In 1986, Jerry Bertrand chaired a symposium in Ontario for four hundred scientists on the future pro-

Great egret

tection of wintering bird habitats, and there were many calls at that conference for more studies on the connection between subtropical habitats and migratory birds, and more active participation in the protection of endangered habitats in the Neotropics. Even before the conference, though, Massachusetts Audubon was at work in the small Central American nation of Belize, which, because of a fluke of economic history and population density, happened to have some 90 percent of its land still undeveloped.

One of the strategies of Harriet Hemenway and company was to provide money to start other bird protection organizations, and then work behind the scenes to see that they became established. This was, in effect, the same technique that Jerry Bertrand employed to do something about the sad decline of forests in the Neotropics. Once again, James Baird, who up until his retirement in 1991 had served as a sort of knight in shining armor in the service of various Arthurian kings of Audubon, sallied forth. Central America was hardly new territory for him. In the 1960s he had represented the Society in pioneering an ecotourism program in the region. Now he again began spending time in Belize.

In 1984, using funds made available by the board of the Massachusetts Audubon Society, Baird started working in association with the local Belize Audubon Society to hire a staff and help them set up management programs for the national parks. Then, in 1987, the board of Massachusetts Audubon again decided to extend the Society's purview: It provided $500,000 to launch the Programme for Belize, a conservation group organized to preserve land, encourage economic development, and advise the government on environmental matters. The aim was to work with Belizeans so that eventually the Programme would be locally run (which it now is). Funds for the Programme were solicited by a consortium of conservation groups spearheaded by Massachusetts Audubon. One of the Programme's main goals was to demonstrate that it is economically feasible to set aside large tracts of land for conservation. The Programme began by acquiring several large tracts of unspoiled forest in northern Belize to hold in trust for the Belizean people; eventually, it created the Rio Bravo Conservation and Management Area, a vast tract of some 230,000 acres that is managed for wildlife conservation and research. The goal is for the Programme and the region's people to be self-supporting, in part from ecotourism in the region, and from the collection and sale of renewable forest products such as chicle.

Another relatively undisturbed and progressive country in Central America is Costa Rica, which in the 1960s designated large tracts of land as national parks—at least in theory. In fact, purchase was intended to come later. Land of course is at a premium in the populated regions of Central and South America, and Bertrand and company realized once more that, unless land were saved outright, soon the great bulldozer that is even now rolling over the tropical regions of the world would squash even the well-intentioned parks of Costa Rica. Two of the largest parks were close to each other but were not contiguous, so Massachusetts Audubon began a mission to connect them and to enhance the wild habitats beyond the formal park boundaries, either through acquisition or by means of nontraditional methods such as conservation easements.

Working with local nongovernmental agencies

and international funding agencies that have an interest in tropical conservation, a Massachusetts Audubon consultant, Andrew Kendall, began putting together the funds and programs to save the tracts of land near the national parks and permanently establish the designated parks.

One of the most interesting of the alternative methods Kendall used to preserve land in Costa Rica involved the U.S. carbon offset project, which is designed to slow the accumulation of gases associated with the greenhouse effect by balancing emission and absorption of carbon in the earth's atmosphere. One of the best ways to ensure adequate absorption of carbon is to preserve the great forest tracts of the planet, which use carbon for photosynthesis. A major contributor to Kendall's effort in Costa Rica was a U.S. power company that, in essence, traded rights to release a certain amount of carbon for money donated to the Massachusetts Audubon-sponsored program to purchase forested parklands in Costa Rica.

THIS COMBINATION OF using innovative international funding strategies, cooperating with the business world, and jumping traditional political boundaries to work in small Central American nations may seem far removed from the Clarendon Street parlor where it all began. But, in fact, one should not suppose that because they dressed in frumpy clothes and preferred a low profile, the families that started the Massachusetts Audubon Society were not powerful manipulators of funds and governmental agencies, with influential connections in the political arena.

The defining characteristic of the Brahmins, the thing that sets them apart from high society in other cities in America, is that they put their power to good use. In our time, as wildlands throughout the world are shrinking, as habitats and species are lost and the human population continues to expand at exponential rates, the little bird club founded by Harriet Hemenway and company back in 1896 could serve as a model for international conservation in the twenty-first century.

Were she with us now, no doubt the *grande dame* would approve.

Rooms in the House of Nature

Northern harrier, Miacomet Plains, Nantucket

A Guide to the
Natural Communities of Massachusetts

*L*eft to their own wanton devices, young children instinctively find what is most obviously attractive in nature: a banquet of sensuous delights. Butterflies and starfish for the eyes; caroling thrushes and booming surf for the ears; skunk and honeysuckle for the nose; the feel of icy brook water and mole's fur; the taste of wild berries and sorrel; each in its place and all blending into a nameless thrill, a foretaste of passion to come. These sensations, tasted in feral idleness, breed curiosity and exploration that may end in learning and even wisdom. This, ideally, is how we should learn about the natural world and our place in it, not from books.

But as a species we are compulsive organizers, constantly assembling what we think we've learned into groups and lists and categories and priorities. And sometimes our obsession with unnatural order—for in nature perfection rhymes with chaos—leads us astray. Purveyors of guidebooks long ago recognized this human weakness for organization and a concomitant tendency to specialize. As a result, people tend to think of nature as *The Wonderful World of Birds*—or butterflies, or orchids—the rest of nature sort of filling in the background. Not that mastery of one aspect of the natural world is bad; in some cases, it is probably a necessary prerequisite to broader understanding and more complex enjoyment. But by itself such a focus distorts the way nature handles her affairs. No doubt it enriches our appreciation of a scarlet tanager to compare it with other spectacular tanagers, but nature never puts all the tanagers together; this happens only in a museum or a book. In nature the tanager is a splendid detail in the canopy of the deciduous forest, one species in a complex community of organisms, some obscure, many as brilliant in their way as the tanager, but all contributing to a magnificent whole that transcends the sum of its parts. As with a great painting, the details of the forest (or bog or barrier beach) may give us great pleasure, but it is the overall "composition" that moves us deeply. Another problem with the stamp-collector approach is that it tends to lure us into one fascinating room of nature, where we become so enthralled that we cease to explore for further treasures and never come to appreciate the encompassing architecture. More than a few promising naturalists have gotten so stuck on birds, for example, that while their life lists grow and their keenness ever sharpens, they remain forever oblivious to the delights of whales and alewife runs—to say nothing of twayblades, hairstreaks, and mudpuppies.

This book is, of course, yet another human exercise in creating artificial order. It does, however, offer an alternative to the taxonomic view of nature, and tries to give a sense of what it's really like out there. We've tried to encourage the reader to view the "nature of Massachusetts" not as a hierarchy of mammals, birds, reptiles, amphibians, fishes, invertebrates, fungi, bacteria, and protozoans, but rather as a realm of intermeshing natural communities. The field marks in this guide are not stripes or spots on particular species, but the species themselves that in characteristic com-

wealth and that spruce-fir forest is concentrated in the western highlands, the communities profiled for the central peneplain are found virtually throughout the Commonwealth, and vestiges of pitch pine–scrub oak barrens—which many of us associate with salt air—still stand in the Connecticut River Valley. The challenge and the rewards of this antitaxonomic approach lie in our perceiving nature's larger designs by recognizing distinctive pieces and seeing how they fit together. Once you have mastered the concept, in an important sense, you will never be lost in the wild places of Massachusetts.

Defining nature is like trying to fathom the cosmos: Certain characteristics and manifestations can be cited, but capturing the complexity, mystery, and splendor of the whole may be beyond human eloquence. Still, we are bound to try, since the more we know of nature—the more we try to encompass her—the grander and more fascinating she becomes.

binations make up the identity of a given community. Some species clinch the identification of a natural community by their presence: If you are lucky enough to happen upon the gorgeous little red and green butterfly called Hessel's hairstreak, you must be standing in an Atlantic white cedar swamp because this butterfly is an *obligate* species, in other words, one that lives only in such swamps. Other species may be strong *indicators*, but not *obligates*: No species is more indicative of a red maple swamp than red maple, yet it can also be quite common in northern hardwood forest. Still other species, such as the black-capped chickadee, live in many natural communities and are part of the biological fingerprint of them all.

In keeping with nature's blithe disregard for tidy categories, we have ordered the chapters to suggest a journey across the state from Stellwagen Bank in the east to Mount Greylock in the west, rather than lumping all forests or wetlands or grasslands into their own neat chapters. Thus you will come to the edge of the oak-conifer forest on p. 96 and encounter a vernal pool on p. 105, just as you might during a spring walk in the woods. The reader should not interpret the relationship between geography and habitat too narrowly, though. While it is true that tide pools are restricted to the eastern extremity of the Common-

Black-capped chickadees

Ocean

Humpback whale and greater shearwaters

The firmaments of air and sea were hardly separable in that all-pervading azure; only the pensive air was transparently pure and soft, with a woman's look, and the robust and man-like sea heaved with long, strong lingering swells, as Sampson's chest in his sleep. Hither and thither, on high, glided the snow-white wings of small, unspeckled birds; these were the gentle thoughts of the feminine air; but to and fro in the deeps, far down in the bottomless blue, rushed mighty Leviathans, sword-fish and sharks; and these were the strong, troubled, murderous thinkings of the masculine sea.

HERMAN MELVILLE
Moby Dick

T he sea holds a unique fascination for our essentially terrestrial species. Its vastness is barely comprehensible. Its power, especially combined with the not always gentle force of wind, is fearsome and continues to claim thousands of human lives every year. Contrary to Melville's metaphor, the sea is also in a very real sense the womb of the earth from which all terrestrial organisms—as well as all aquatic and marine ones—originated. In fact, over millennia some life forms have come out onto the beach, evolved into land creatures, and then gone back into the sea. Yet despite what it has given up, the ocean still contains a quarter of a million species.

Although the climate, and therefore the natural history, of Massachusetts is influenced by global ocean phenomena such as the El Niño current off the west coast of South America, for the purposes of this book it is practical to define a more limited "Massachusetts ocean." The Gulf of Maine serves this purpose well. This roughly rectangular depression of 36,000 square miles is bounded by the New England coast to the west; Georges Bank and Browns Bank to the east; Cape Sable, Nova Scotia, to the north; and Nantucket Shoals to the south. The Gulf's natural boundaries give it a high degree of biological identity and encompass the strong warm-water influences that prevail south of Cape Cod as well as the cold-water sea that fills the basin to the north. The general description that follows attempts to put our local ocean in a global context; indicator species and other particulars relate specifically to the Gulf of Maine.

Area in Square Miles
Gulf of Maine: 36,000
Land area of New England states: 66,608
Atlantic Ocean, Greenland to Antarctica: 31,529,000
Global ocean: 141,254,000 (70.8 percent of the earth's surface)

Depth
Average depth of the Gulf of Maine: 492 feet (82 fathoms).
Maximum depth of the Gulf of Maine: 1,236 feet (206 fathoms), in Georges Basin; there are twenty other deep basins.
Depth of Georges Bank: Mostly less than 200 feet, with shoals to within 13 feet of the surface.
Depth of Stellwagen Bank: Shallowest areas, 71.5–108 feet.
Average depth of the North Atlantic: 12,880 feet.
The only deepwater connection between the Gulf of Maine depression and the Atlantic Ocean is the 762-foot-deep Northeast Channel, which separates Georges Bank from Browns Bank.
Greatest ocean depth recorded: The Marianas Trench, 36,160 feet under the surface of the Pacific. (If you dropped Mount Everest into this trench you would have to swim down more than a mile to reach its summit.)
The force of the water's weight increases as you descend into the ocean depths. At 3,000 feet the pressure is enough to compress a block of wood to about half its volume. Thus, we may reasonably be impressed that sperm whales regularly sound to

3,000 feet, are known to reach depths of over 7,000 feet, and may well descend to 10,000 feet.

Wind and Waves

The aspect of the sea to which we pay the greatest attention as we set out on it, seasickness remedies at hand, is its surface motion. A passing boat may create a wake, or an undersea earthquake or volcanic eruption may set in motion a tidal wave, or "tsunami." For the most part, though, the charac-ter of the sea surface is created by the wind. Without the effects of wind, the sea is nearly flat and glassy—a rare occurrence in New England waters. But the slightest breeze puts the ocean in motion. This motion involves little actual forward movement of water, but rather the movement of a force, a wave of energy, which lifts the sea surface into a series of ripples or a mountain range of towering seas.

The size of wind-driven waves is determined by

three factors: wind speed; the duration of a blow; and the distance the wind travels unimpeded in the same direction over the ocean surface, called fetch. The biggest waves are generated by all of these factors working together: strong winds blowing for a long time over a wide expanse of sea. The creation of waves is a cumulative effect. The smallest ripple presents a surface for the wind to push against, which tends to make it bigger. A law of physics says that when a wave's height reaches

one seventh its length (the distance between the crests of successive waves), it will break and add its force to a higher wave with a longer wavelength. As a storm comes on, you can watch the wind blow smaller whitecaps together to form ever taller, more widely spaced waves. These larger, more stable waves are called seas. The waves that continue to travel after the wind drops are called swells. Waves may travel across entire oceans. It is not unusual to see a crashing surf along the Massachusetts coast on a clear, calm day. This may result from a storm that passed hundreds of miles away. A powerful storm blowing for several days over a long fetch can generate very large waves, whose height tends to be about half the wind speed. For example, a hurricane with 80-mile-per-hour winds can be expected to generate 40-foot waves at sea. The highest ocean storm wave ever reliably recorded occurred in the central Pacific and crested at 112 feet. Experienced mariners can judge the speed of the wind simply by looking at what it is doing to the sea surface.

As waves approach the shore, they "stub their toes" on the bottom and break. The ratio at which waves break is 3:4, that is, when a 3-foot wave reaches 4-foot depth, the friction with the sea bottom slows the wave down, which narrows the interval between the waves behind as they catch up. This force increases the height of the front wave and forces the peak forward. If a beach has a gradual slope out to sea, the waves tend to spill forward relatively gently; if the shore has a steep dropoff and the waves hit bottom abruptly, they will rise more steeply and plunge more violently. Surfers are intimately familiar with the difference between "spillers" and "plungers."

In addition to making life interesting for surface-feeding animals and those that live along a rocky shore, waves aerate the water, providing necessary oxygen for fish and other organisms, and at shallow depths help bring nutrients to the surface.

Temperature

The Gulf of Maine is a cold sea, containing northern, or boreal, organisms that become scarce or absent as the influence of the Gulf Stream warms the ocean to the south of Cape Cod. The average surface temperature in the Gulf of Maine in August ranges between 59°F and 68°F (warmer in the coastal shallows, but not much!); the average February sea surface temperature off New England is between 32°F and 41°F. Worldwide, the highest recorded sea surface temperature is 96.8°F inshore in the Persian Gulf in the summer.

Although in the summer in New England the sea is normally cooler than the atmosphere, in the winter it is much warmer on average, so overall the Massachusetts ocean is a warmer medium than the air. The constantly changing relationship between ocean and air temperatures strongly influences coastal weather patterns.

The sea's temperature varies with depth, latitude, and season, and these variations profoundly influence the abundance and distribution of marine life. In boreal seas like the Gulf of Maine there is a seasonal turnover, just as there is in a freshwater pond or lake. In summer the sea's surface water is relatively warm and light and the bottom layer, cold and dense; the extremes are separated by a zone of steeply decreasing temperature called the *thermocline.* As the surface water cools and becomes heavier in the autumn, the layers begin to mix, the thermocline breaks, and surface and bottom temperatures turn over. By contrast, tropical seas have a very stable water column with no turnover because the surface temperature remains essentially unchanged, an average of 82°F year-round at the Equator, as does the cold temperature in the depths.

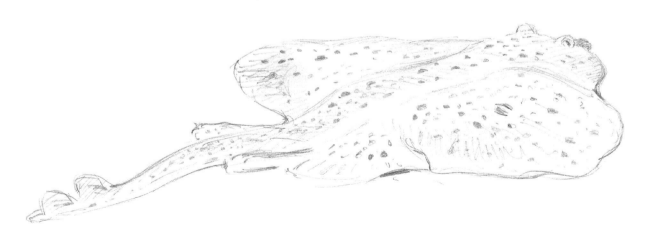

Big skate

The mixing of bottom and surface waters in temperate-zone seas brings nutrients to the surface that are a food source for a wealth of minute, plantlike organisms, called phytoplankton, which drift near the surface; these are preyed upon in turn by the zooplankton and so on out through the food web to fishes, seabirds, and great whales. This explains why the colder oceans are murky and support an abundance of life, while the crystal-clear surface waters of the tropics are typically rather sterile. In addition to seasonal mixing, cooler water and bottom nutrients are also brought to the surface by upwellings that occur where currents are pushed upward against geological barriers such as the walls of submerged banks or the edges of continents. Rich concentrations of fish, birds, and mammals are often associated with these upwellings, as on Georges and Stellwagen banks.

Salinity

Why is the sea salty? The sea is like a great sink constantly receiving tons of minerals from precipitation and the consequent erosion of the earth's surface. The minerals in seawater come partly from this runoff and partly from dissolved minerals from the seabed itself. Some scientists believe that the sea has always been salty, owing to the chemical solution that was created as the earth and its oceans were forming and the chemicals contained in the earth's earlier atmosphere, which reached the ocean in rain and snow. These chemicals have then combined and reacted in different ways over the eons, some falling out of solution into the sea floor and some evaporating from the surface.

Seawater has an average salinity of 3.5 percent, meaning there are 35 parts of salt to every 1,000 parts of seawater, by weight. This is not inordinately salty by earthly standards; the salinity of the Great Salt Lake, for example, ranges between 5 and 27 percent. Salinity levels in the Gulf of Maine vary from place to place. Two hundred fifty billion gallons of fresh water a year flow into the Gulf, mainly from rivers in Maine, creating low salinities along its western edge, and elsewhere by circulation and currents. The average salinity of 3.3 percent in the Gulf of Maine is on the low side, compared to the open ocean.

The presence of salt lowers the freezing point of seawater. At a salinity of 3.5 percent the ocean does not freeze until it reaches 29°F, 3 degrees below the freezing point of pure water.

Tides

Tides result from the gravitational pull of the moon and the sun. Because it is much closer, the moon exerts 2¼ times more gravitational pull on the earth's oceans than the sun, despite the latter's far greater size. High tides occur both on the side of the earth directly facing the moon and on the opposite side of the earth, since the moon's gravity also, so to speak, pulls the earth away from the oceans on the far side. Imagine a bulge of saltwater on opposite sides of the earth in line with the moon. When the sun and the moon are both in line with the earth, the extra pull exerted by the sun creates extreme high and low tides, called spring tides. This event has nothing to do with the season, but occurs at the full and new moons. When the sun and moon are at right angles in relation to the earth (first-quarter and third-quarter

Pollock

moons) the tides have the least range between high and low. These are called neap tides. Other factors, including the shape of coastal features, wind, and barometric pressure, also affect the height and range of tides.

The tidal ranges in the northern end of the Gulf of Maine are some of the greatest in the world; they decrease as you travel south along the New England coast: In the Bay of Fundy the tidal range is 50 feet; at Passamaquoddy Bay, Maine, 28 feet; in Massachusetts north of Cape Cod, 7 to 10 feet, and south of Cape Cod, 4 feet or less.

Currents

The oceans move constantly in patterns, called currents, that can be likened to great rivers or whirlpools. In fact, the word "ocean" derives from the name of a Greek god Okeanos, a Titan who ruled over a great river encircling the earth. Major currents are caused by the combined effects of the sun, wind, and rotation of the earth.

Because of constant exposure to the sun, tropical oceans are warmer than those near the poles. Warm water is lighter (less dense) than cold water, with the result that the sea level in the equatorial seas is higher than in the temperate zone. Consequently, there is a "downhill" flow of warm water north and south from the Equator and a corresponding movement of cold water below the warm flow, moving in the opposite direction, from the poles toward the Equator.

Another cause of ocean currents is the rotation of the planet. As the earth turns east on its axis the oceans are pushed westward. Because the earth is a rotating sphere, this motion is straight only at the Equator. In the Northern Hemisphere it bends to the right (clockwise) and in the Southern Hemisphere to the left (counterclockwise). This is called the Coriolis effect.

The earth's rotation also sets up a pattern of winds, the reliable trade winds, which blow from northeast to southwest diagonally toward the Equator in tropical latitudes. Thus, the basic North Atlantic current may be thought of as a circular river, or gyre, flowing clockwise. The stretch of this river with which we are most familiar is the Gulf Stream.

New England inshore waters north of Cape Cod are generally influenced by cold boreal currents. The Gulf Stream is deflected eastward by the effects of Georges Bank and normally does not reach within 200 miles of the seaward edge of the Gulf of Maine. However, occasional shifts in the Stream or unusual weather conditions do push warmer water onto Georges Bank, bringing rare fish and birds within range of New England fishermen and pelagic birders—and also creating havoc with the normal fish populations of the bank. Even in average years, the Gulf Stream influence tends to make water temperatures warmer on the south shores of Cape Cod and the islands, and it is not unusual for warm "core rings" to split off from the main stream and carry tropical fish close to these outposts.

The Gulf of Maine has its own gyre that rotates counterclockwise under the influence of three main factors: (1) spring flooding from its rivers, including the powerful spring discharge of the St. Lawrence River, which pushes through Cabot Strait along the east coast of Nova Scotia and into the Gulf; (2) the ebb and flow of strong tides, which in the Northern Hemisphere cause a counterclockwise circulation; and (3) deepwater currents moving southwest along the continental shelf, which swing forcefully into the Gulf through the Northeast Channel, between Browns and Georges banks. The gyre is strongest in spring, reaching a top speed of about 8 miles a day. A cork floating in this gyre would take at least three months to travel around the gulf. This current pattern should be thought of as an average over the course of a year; on a day-to-day basis, especially in summer, fall, and winter, it is extremely variable and its courses and directions unpredictable.

Light

Life on earth is closely linked with light. This is mainly because green plants need light for photosynthesis, herbivores eat plants, and carnivores eat herbivores. A few species have evolved ways of getting along without light, or of supplying their own, but not many. Though water is transparent, it does absorb light, so that the deeper you dive, the darker it gets. Even in the very clearest seas, you will find yourself in total darkness below 400 to 500 feet. In turbid boreal waters the lights may go out at around 30 feet. Available light strongly influences what lives where. In clear tropical waters, enormously diverse coral reefs bask in nearly constant sunshine at up to 80 feet, though nothing much lives in the nutrient-poor water above. In the Gulf of Maine, by contrast, much of the action is at or near the surface, where a rich soup of phytoplankton redoubles itself by "eating daylight," and only a few odd-looking specialists can cope with life in the stressful deeps. These depend for sustenance on organic matter that sifts down on them from above like fish flakes in a vast aquarium.

Ocean Biogeography

All of the elements just described influence how the sea is colonized by marine plants and animals. Given that most of the sea bottom is cold, utterly dark, and crushed under the immense weight of the water above it, it is surprising at first to learn that 98 percent of the 250,000 known species that live in the sea are *benthic* creatures, or bottom dwellers. Many of these creatures live in coastal shallows or coral reefs penetrated constantly by life-giving sunlight. But our ability to plumb the depths with submersible research vessels has revealed that the profundal depths are by no means the uninhabitable wastes that we have imagined, but fairly throb with improbable life forms employing sophisticated lights, breathing devices, and feeding strategies to survive in the abysmal depths.

Beyond the coastal zone in the Gulf of Maine, the greatest concentrations of marine life are at or near the surface—especially where upwellings of nutrients act as a kind of biological magnet for plankton, fish, birds, and whales feeding variously on each other—and on the bottoms of offshore shallows such as Georges and Stellwagen banks.

INDICATOR SPECIES

In general only groups and species that are subtidal, that is, characteristic of the realm between the low tide line and the deep ocean are listed; coastal and deepwater species may be included if they are particularly abundant or characteristic offshore in the Gulf of Maine. Southern species that stray into the southern end of the Gulf are omitted.

Phytoplankton and *zooplankton* are not taxonomic categories but describe the behavior of many species that drift (from the Greek *planktos,* "drifting") in the water column. *Plankton* describes the whole mass; *plankters* are the individual organisms. The taxonomic listing begins with the seaweeds.

Phytoplankton, or "plant drifters," are single-celled algae of many species that contain chlorophyll and thus can use sunlight for photosynthesis. They make up the greatest part of the sea's plankton mass and are the primary producers in the marine food web, converting inorganic elements such as nitrogen and phosphorus into a food source for other species. They are most abundant in early spring, when the greatest amounts of nitrogen and other minerals are available in surface waters.

The most abundant of these algae are the *diatoms,* which come in an astonishing variety of forms: stars, disks, boxes, needles, beads, and many others. Under magnification the hard silica shells (cell walls) of diatoms resemble geometric sculpture or intricate pieces of machinery. Diatoms reproduce mainly by splitting in two (binary fission), but also reproduce sexually.

Another important group of plankters is the *dinoflagellates.* Because they use a long whiplike "tail," or flagellum, to propel themselves in a whirling motion and in many cases also contain chlorophyll, their proper classification has been much debated. They are now generally placed with neither plants nor animals but in the kingdom

Protista. Dinoflagellates also come in many intricate, peculiar, and beautiful shapes. A number of species exhibit bioluminescence—a kind of living glitter that highlights the waves at night and even flashes from the skin of swimmers when these protists "bloom" in large numbers. One type of diatom (*Alexandrium* species) is the organism responsible for so-called red tides and resultant paralytic shellfish poisoning.

Zooplankton, or "animal drifters," often are the early stages of mollusks, crustaceans, and other bottom-anchored creatures. These float and drift for a period, then transform into a heavier stage, sink to the bottom, and walk to a suitable permanent home. But other zooplankters pass their entire life cycles adrift on ocean currents. Some are quite large—for example, jellyfish—but the majority are tiny (1–5 mm) shrimplike crustaceans called copepods. It is impossible to overstate the importance of these minute herbivores in sustaining marine life in temperate seas. They spend their days grazing on phytoplankton and are in turn a primary food source for larger predaceous animals, from fish to birds to whales. If living seas are considered crucial to life on earth as we know it, the nearly invisible copepods are arguably the most important animals on the planet.

❧ **MARINE ALGAE (SEAWEEDS)** are scarce in the deepest waters; those that grow in shallow areas or are seen floating on the surface are among the species found along the rocky shore. The most common seaweeds seen offshore are the rockweeds (*Fucus*), which are detached from coastal rocks by heavy surf and float at the surface, often in long windblown skeins, by means of air bladders in their "leaves." ❧ **EELGRASS:** One of only two seed-bearing plants growing below the low tide line in New England; forms extensive beds in sheltered areas, supplying food and habitat for a number of organisms such as brant, small fish, bay scallops, and other invertebrates. ❧ **SPONGES:** Fig sponge (grows on large bivalves). Many lower intertidal sponges range into subtidal zone. ❧ **HYDROZOANS:** Many species occur as tiny jellyfish (hydromedusae) in plankton. ❧ **JELLYFISH:** Lion's mane (or red) jelly (the largest jellyfish in the world, often 3 feet in diameter; the record is 6.5 feet), moon jelly. ❧ **CORALS:** Star coral (Cape Cod). ❧ **SEA ANEMONES:** Burrowing anemones (at least one subtidal species); cerianatherian anemones (one subtidal Gulf of Maine species); frilled sea anemone, our most common species, ranging into the intertidal zone especially when young. ❧ **COMB JELLIES:** Occur as plankters, sometimes

Sooty shearwater

in great swarms, and are predaceous on small fish, etc. Sea gooseberry, common northern comb jelly, Beroe's comb jelly. ❧ **BRYOZOANS:** Several species of panpipe bryozoans, shelled bryozoan, some subtidal species of lacy crusts, shielded bryozoan, some subtidal species of red crusts, several species of smittinid crusts. Many intertidal-to-subtidal species. ❧ **ARROW WORMS:** Three *Sagitta* species; predaceous plankters. ❧ **LAMP SHELLS (BRACHIOPODS):** Northern lamp shell. ❧ **MOLLUSKS:** Bee chiton, chink shell, alternate bittium (on eelgrass), Greenland wentletrap, shelled sea butterflies (*Limacina retroversa* is the only Gulf of Maine species; plankter), naked sea butterfly, Sacoglossan slug (on eelgrass), nudibranchs (about 12 species, mostly intertidal-to-subtidal), ax yolida, file yolida, short yolida, oval yolida, horse mussel, deep-sea scallop, mahogany (or black) clam, chestnut astarte, northern cardita, Greenland cockle, surf clam (mainly subtidal), Arctic wedge clam, dwarf tellin, chalky macoma, Conrad's thracia, paper spoon shell, Lea's spoon shell, boreal squid (predaceous on pelagic fish), offshore octopus. ❧ **POLYCHAETE WORMS:** An enormously diverse group with hundreds of species in this region. Plankton worms, sea mouse, chevron worm, plus many intertidal-to-subtidal species, including the common fishing-bait clam worms or bloodworms. ❧ **SIPUNCULAN WORMS:** Hermit sipunculid. ❧ **INSECTS:** There is only one truly sea-going insect, a water strider that skates over the usually calm waters of the Caribbean. However, a number of highly migratory insects are regularly seen flying offshore in New England, notably monarch, painted lady, red admiral, and occasionally other butterflies and the wandering glider, the only cosmopolitan dragonfly. ❧ **CRUSTACEANS:** Crenate barnacle, striped goose barnacle, several species of pelagic goose barnacles, 2 species of big-eyed amphipods, several species of calanoid copepods, planktonic amphipod, horned krill (most common euphausiid in the Gulf of Maine), boreal red shrimps (3 *Pandalus* species are the edible "Maine" shrimps), northern lobster, hairy hermit crab, Acadian hermit crab, *Lithodes maia* (crab), toad crab; the common rock and Jonah crabs are mainly subtidal. ❧ **ECHINODERMS:** Silky cucumber, other sea cucumbers are

Northern lobster

mainly intertidal-to-subtidal, sand dollar, mud star, spiny sunstar, purple sunstar, blood star (mainly subtidal), basket star. Most urchins and sea stars range into deep water. ~ **TUNICATES:** Sea pork, stalked sea squirt (other sea squirts range from intertidal into subtidal and deep water), sea peach (3 *Molgula* species), appendicularians (2 *Oikopleura* species; minute plankters). ~ **FISHES:** Marine fishes are by far the most diverse class of organisms on earth, with over 20,000 known species. Most of these are tropical. Bigelow and Schroeder (1953) recorded a mere 211 species in the Gulf of Maine, most of which are rare in our waters. Species that occur regularly today include: Sea lamprey (A), sand shark, blue shark, mackerel shark, basking shark (the largest fish in New England waters, at up to 50 feet and 8,600 pounds), smooth dogfish and spiny dogfish, barn-door skate, big (or winter) skate, little skate, smooth-tailed skate, thorny skate, Atlantic sturgeon (A, SE), short-nosed sturgeon (A, FE), Atlantic herring, menhaden, American shad (A), blueback herring (A), alewife (A), Atlantic salmon (A), rainbow smelt (A), American eel (C), silver hake, Atlantic cod, haddock, pollock, white hake, red hake, cusk, Atlantic halibut, American plaice, yellowtail flounder, winter flounder, gray sole (or witch flounder, windowpane (or sand flounder), Atlantic mackerel, bluefin tuna, swordfish, butterfish, bluefish, striped bass (A), porgy (or scup), redfish (or rosefish), grubby, shorthorn sculpin,

longhorn sculpin, sea raven, lumpfish, cunner, tautog, sand lance, rock gunnel, ocean pout, ocean sunfish, American goosefish (monkfish). ~ **REPTILES:** All of the sea turtles breed on tropical beaches and all but the leatherback occur in the Gulf of Maine as stranded animals that become cold-stunned and trapped in Cape Cod Bay. Sea turtles: green turtle (FT), hawksbill (FE), Kemp's (Atlantic), ridley (FE), loggerhead (FT), leatherback (FE; largest North American reptile). ~ **BIRDS:** Most of our true pelagic birds are seasonal visitors to the Gulf of Maine, either arriving for the winter from northern breeding grounds or en route, migrating to and from Arctic or Antarctic breeding grounds. Common loon (SSC, M), northern fulmar, Cory's shearwater, greater shearwater, sooty shearwater, Manx shearwater (one nesting record in Massachusetts), Wilson's storm-petrel, Leach's storm-petrel (SE, M; a small colony breeds on Penikese Island), northern gannet, red-necked phalarope, red phalarope, pomarine jaeger, parasitic jaeger, long-tailed jaeger (rare), herring gull (M), great black-backed gull (M), Iceland gull, glaucous gull, black-legged kittiwake, common tern (SSC, M), Arctic tern (SSC, M), dovekie, common murre (rare), thick-billed murre, razorbill, Atlantic puffin (rare in Massachusetts). ~ **MAMMALS:** Vagrant species not typical of the ecosystem are omitted. Sperm whale (FE; rare, occurring mainly in deep water beyond the continental shelf), bottle-nosed dolphin, white-beaked dolphin, Atlantic white-sided dolphin, killer whale (rare), common pilot whale, harbor porpoise (candidate for FT), fin whale (FE), sei whale (FE), minke whale, blue whale (FE; rare), humpback whale (FE), northern right whale

(FE; possibly the most endangered of the world's great whales), harbor seal (winter resident in Massachusetts), gray seal (pups in midwinter on islands in Nantucket Sound).

CONSERVATION STATUS

Given what the sea has done for us, it is a little depressing to contemplate how we have treated her in return. In less than 400 years we have

- Overharvested commercial fish stocks to a point at which the industry is on the brink of collapse. Georges Bank, once one of the richest fisheries in the world, is now depleted of many species and has been partially closed by federal authorities in hopes that stocks can recover.
- Made six of our great whale species globally endangered, possibly pushing the northern right whale beyond the point of no return. The harbor porpoise, a common species here until recent decades, is now a candidate for federal listing as threatened and continues to be trapped and drowned in commercial gill nets.
- Grossly polluted coastal waters with heavy metals, PCBs (polychlorinated biphenyls) and other toxic industrial-waste products, oil from spills and routine bilge pumping, and a stupendous assortment and volume of nonbiodegradable trash.
- Overcrowded coastal waters with recreational boats, which may be a factor in the continuing decline of the northern right whale.
- Created ecological imbalances such as the population explosion of great black-backed and herring gulls. Ocean dumping of fish wastes in a vastly expanded commercial fishery and the open disposal of more and more human-generated garbage at megadumps greatly increased the productivity of these species. The gull boom in turn threatens populations of smaller, less aggressive migratory species such as terns and piping plovers.

On the bright side, we have identified many of the environmental issues, and many of the solutions as well. All of them cost money to implement, and most restrict the behavior of various special interest groups. It remains to be seen whether we will have the political will to turn the tide of chronic abuse that we have visited on the mother of all ecosystems.

PLACES TO VISIT

The best way to experience the ocean and its wildlife is to go out on it. In Massachusetts there are now whale-watching trips sailing from Newburyport, Gloucester, Boston, Plymouth, Provincetown, and Nantucket. Most of these boats have competent naturalists on board, and many collect scientific data on the whales observed. Typically, full- and half-day trips go to Stellwagen Bank, Jeffreys Ledge (north of Cape Cod and Cape Ann, respectively), or to Nantucket Shoals; even though whale populations change feeding grounds unpredictably from year to year, few cruises come home without a sighting. Some operators and birding groups do overnight cruises to Georges Bank and the continental shelf. Organizations such as the Massachusetts Audubon Society and the New England Aquarium, and some cruise operators, offer interpretation of the marine ecosystem as a whole and don't focus merely on whale spotting.

Striped bass

FURTHER READING

Exploring the Oceans, by Henry S. Parker. Englewood Cliffs, N.J.: Prentice-Hall, 1985.

A Field Guide to Whales, Porpoises and Seals from Cape Cod to Newfoundland, by Steven K. Katona, Valerie Rough, and David T. Richardson. Washington, D.C.: Smithsonian Institution Press, 1993.

Fishes of the Gulf of Maine, by Henry B. Bigelow and William C. Schroeder. Washington, D.C.: U.S. Fish and Wildlife Service, 1953.

The Gulf of Maine, by Spencer Apollonio. Rockland, Me.: Courier of Maine, 1979.

Marine Ecological Processes, by Ivan Valiela. New York: Springer-Verlag, 1984.

Moby Dick, by Herman Melville.

The Open Sea: Its Natural History, by A. Hardy. Boston: Houghton Mifflin, 1965.

The Sea Around Us, by Rachel L. Carson. New York: Oxford University Press, 1950.

Stellwagen Bank: A Guide to the Whales, Sea Birds, and Marine Life of the Stellwagen Bank National Marine Sanctuary, by Nathalie Ward. Camden, Me.: Down East Books, 1995.

Whalewatchers' Guide to the North Atlantic, Lincoln, Mass.: Massachusetts Audubon Society, 1989. Pocket field guide.

Rocky Shore and Intertidal Zone

The pools are gardens of color composed of the delicate greens and ocher-yellow of encrusting sponge, the pale pink of hydroids that stand like clusters of fragile spring flowers, the bronze and electric blue gleams of the Irish moss, the old rose beauty of the coralline algae.

And over it all there is the smell of low tide, compounded of the faint, pervasive smell of worms and snails and jellyfish and crabs—the sulphur smell of sponge, the iodine smell of rockweed, and the salt smell of the rime that glitters on the sun-dried rocks.

RACHEL CARSON
The Edge of the Sea

Purple sandpipers

O n the North Shore of Massachusetts a grand crumbling seawall of Ordovician granite stretches from the low skerries of Boston Harbor to the pink promontories of Cape Ann. After a hundred miles or so of shifting coastal sands, the wall reappears just south of Portland as the "rockbound coast of Maine" and then keeps going, with only brief interruptions, up into the Arctic, over to Greenland and Iceland, and down through the British Isles without pausing for sandy interludes of any scope until France. The segment of this rocky edge that runs between the frigid waters north of Labrador and the comparatively tepid seas south of Cape Cod lies within the American Atlantic Boreal Region, a distinct province of marine life with its own characteristic organisms. Although many of these marine plants and animals are seabed creatures that reach the landwardmost extent of their distribution in the rocky intertidal community along the shoreline, others prefer this wave-washed rocky habitat over any other.

The unconsolidated and constantly shifting nature of the intertidal zone of a sandy beach makes it a tough place to settle permanently: Except for a few burrowers, the transient shorebirds, and the fair-weather naked apes, a sandy beach between the tide lines is not a very lively place. The rocky intertidal zone, by contrast, is typically so crowded with plants and animals that it may be hard to locate a patch of uncolonized bare rock. The reasons for this striking difference are quite obvious. Rock provides a firm surface to which organisms can attach themselves securely, and this kind of shore also contains an infinity of hospitable irregulari-ties, especially pools of varying depths, temperatures, degrees of shelter, and chemical flavors, where even the most finicky alga or crustacean can make a living—if it can tolerate the twice-daily pounding of waves. Furthermore, the tides clean house and deliver fresh food twice a day.

Like crowded human communities, this high-density habitat is a battleground in a continuous turf war, crustacean challenging seaweed and mollusk pressing polyp for a piece of the rock. The rocky intertidal zone may be divided into a number of more or less distinct subzones. The *spray zone*, which tastes salt only from the air or in the highest storm tides, is mainly bare rock. Where a little soil has accumulated in cracks, you are likely to find some of the same hardy plants that you find in sand dunes, such as seaside goldenrod or beach pea. In addition there may be patches of yellow or orange lichens in the genera *Xanthoria* and *Calo-placa*. The shore, or yellow wall, lichen (*Xanthoria parietina*) is also common on stones in seaside cemeteries. *Caloplaca* species often indicate the presence of seabirds or, more precisely, their droppings.

The *black zone*, just below the high-tide line, is named for a dense mat of dark blue-green cyanobacteria (kingdom Monera), once thought to be a kind of seaweed; the mat is composed of microscopic individual plants of several species, mostly in the genus *Calothrix*. This slippery composite fixes itself to the rocks and retains moisture by secreting a gluey substance. This highest tidal zone remains out of the water the longest time in any tide cycle, and few other marine organisms can withstand such prolonged immersion in the

air. Another exception is the rough periwinkle, which can survive for as long as a month out of water and regularly grazes on this black pasture.

The next strip down is the *periwinkle-barnacle zone*, named for its dominant animals. The mobile periwinkles cannot always be counted upon to stick to their appointed zone, but the white barnacles are permanently attached and stand out as a bright ribbon between the black zone above and the skeins of seaweed below.

The *rockweed*, or *brown seaweed, zone* is composed of several luxuriant yellow-brown seaweeds in the genera *Fucus* and *Ascophyllum*, often called rockweed, bladder weed, or wrack. Like a terrestrial forest, these seaweed communities provide food and cover for many other intertidal organisms. Blue mussels may be common or even dominant across both the periwinkle-barnacle and rockweed zones.

The *Irish moss zone* is the lowest intertidal horizon and harbors an abundance of life in its dense thickets. It is easily recognized by the presence of the short, tufted, reddish seaweed for which it is named. The species shows a lovely bluish iridescence underwater and can grow at greater depths than some other seaweeds because of its light-trapping capability. On vertical rocks it is exposed only during the lowest of monthly spring tides, but it also grows in the lowest tide pools.

The *kelp zone* has only a tenuous claim to inclusion within the rocky intertidal community. The sea drops to this level only twice a year, during the extreme equinoctial tides. The long brown ribbons of kelp attached by tangled, rootlike "holdfasts" are unmistakable. Most of the marine life that dwells amid the kelp has little tolerance for

Halibut Point, Rockport

life out of water and is best observed using snorkel and mask or scuba gear.

A breakwater shows these zones most clearly, but the most rewarding intertidal explorations are done at low tide along a rocky shore with plenty of tide pools. In general, the lower you go the more interesting things get. In the brackish, often rather unsavory-looking pools most remote from the ocean, you may see nothing but the brilliant green strands of mermaid hair seaweed, whose scientific name, *Enteromorpha,* refers to the intestinelike form of its leaves. Deep pools near the low tide line, by contrast, are often crammed with colorful marine forms such as anemones and sea urchins. Because they are covered by the sea most of the time, these low pools often contain creatures characteristic of the kelp zone or ocean floor.

The rocky shore is alluring at all seasons, even when the weather is "bad." The colors of sea, sky, and rock are ever-changing: Rocky headlands have been favorite subjects of New England landscape painters such as Childe Hassam, William Trost Richards, and FitzHugh Lane. The wide variety of strange life cycles enacted by marine organisms creates constantly shifting scenes in the tide pools. During the winter many species of seabirds feed along the rocky edge. And there is no greater outdoor thrill in Massachusetts than standing at a respectful distance and feeling the crash of immense storm waves as they pound into the granite wall below.

INDICATOR SPECIES

Some of the organisms listed live mainly below the low tide line but commonly appear in the lowest tide pools or are washed up into the intertidal zone on rocky shore beaches. ∾ WILDFLOWERS (spray zone): Beach pea, seaside plantain, seaside goldenrod, knotted pearlwort (ST). ∾ LICHENS: Shore lichen, other yellow and orange species of the genera *Xanthoria* and *Caloplaca.* ∾ MARINE BACTERIA: *Calothrix* species, once considered seaweeds, form the dark, slippery surface on coastal rocks from which the black zone takes its name. ∾ MARINE ALGAE (SEAWEEDS): Mermaid hair; sea lettuce; four common rockweeds, or wrack, species in the genera *Fucus* and *Ascophyllum*; Irish moss; tufted redweed; laver (the edible Nori prized in Japan); dulse; coral weed; species of crustose seaweeds, a number of which make bright pink or red patches on the walls of tide pools. ∾ INSECTS: Tide pool insect, a springtail, is the only cold-water marine insect. ∾ SPONGES: Crumb-of-bread sponge, finger sponge. ∾ CNIDARIANS: Club hydroid, pink-hearted hydroid, snail fur (colonizes the surface of a snail shell inhabited by a hermit crab), common or frilled sea anemone. ∾ WORMS: Coiled worms are tiny white snaillike coils attached to seaweeds. ∾ CRUSTACEANS: Northern rock barnacle (O), rock crab, Jonah crab, green crab (X), scuds and gammaridian amphipods. ∾ MOLLUSKS: Blue mussel, red chiton, common periwinkle (X), rough periwinkle, smooth, or yellow, periwinkle, dog whelk, tortoise-shell limpet. ∾ NUDIBRANCHS (shell-less mollusks): Masked, or plumed, sea slug, red-gilled sea slugs, rough-mantled sea slugs, bushy-backed sea slug. ∾ ECHINODERMS: Common (or Forbes's) sea star (starfish), purple sea star, daisy brittle star, dwarf brittle star, green sea urchin, sea cucumber. ∾ BRYOZOANS, OR MOSS ANIMALS: Tufted bryozoan, sea lace. ∾ TUNICATES: Sea

KEY
O=Obligate
ST=State Threatened
X=Exotic

Double-crested cormorant

Purple and common sea stars

pork, sea vase, star tunicate, sea potato, sea grape. ∾ **FISHES:** Rock eel, radiated shanny, sea snail, fourspine stickleback, cunner, tautog, striped bass. ∾ **BIRDS:** Two bird species, both winter visitors from farther north, are essentially restricted to the rocky shore: the exquisite harlequin duck and the purple sandpiper; the black guillemot is also strongly associated with this habitat. Other birds that are characteristic of, but not restricted to, the rocky shore include common loon, horned grebe, red-necked grebe, great cormorant, double-crested cormorant, common eider, king eider, black scoter, surf scoter, white-winged scoter, common and Barrow's goldeneyes, bufflehead, red-breasted merganser, herring gull, great black-backed gull, and Iceland gull. Many of our sandpipers and plovers feed in the rocky intertidal zone and roost on the rocks at high tide; the ruddy turnstone and red knot are particularly fond of this habitat. The arctic-breeding snow bunting is the only songbird that might be considered at home on coastal rocks. Most of the birds noted above occur in New England mainly between November and April and breed to the North. ∾ **MAMMALS:** Harbor seal. The sea mink, a large weasel, was reported to have foraged along the shore as late as the colonial period, but was extirpated soon after European settlement.

DISTRIBUTION

Rockbound coasts are comparatively scarce and localized worldwide, compared to sandy shores. They occur on all continents where intrusions of magma have erupted from the seabed. Of course, limestone, metamorphic, and sedimentary formations along the coast are also "rocky," though they may support life forms quite different from those treated here. The American Atlantic Boreal Rocky Shore runs from just south of Boston north to Labrador.

EXTENT IN MASSACHUSETTS

Rocky shore accounts for approximately 50 miles of the Massachusetts coast, a figure that would be much higher if all the many coastal irregularities and the perimeters of all our rocky islands were accurately measured. It is a decidedly narrow, linear community, rarely spanning more than 200 feet between the lowest tide line and the upland edge of the spray zone.

CONSERVATION STATUS

Because the intertidal zone is so dynamic and ever-changing, it is difficult to study the status of the organisms that live there. If they are present one year and absent the next, who can say which of dozens of variables is responsible? Although there is a growing sense among some biologists who have studied the rocky shore for years that diversity is declining and that certain species (e.g., nudibranchs) long common are becoming scarce, the problems of documenting possible declines scientifically remain. No species of marine invertebrate appears on the Massachusetts Endangered Species list, one of the most extensive in the country. Whether this results from lack of rarities or lack of information is hard to say.

In Europe and Asia, rocky shore invertebrates such as blue mussels, periwinkles, and sea urchins are gathered routinely by visitors to the shore much as we gather blueberries. Both casual and professional harvest of this kind is on the rise as is aquaculture involving some species.

Several edible seaweeds occur in abundance along our rocky shores, and quantities of Irish moss and other edible species are harvested in some areas. Rockweeds are also gathered in large quantities for clambakes and fertilizer. While seaweeds seem to regenerate readily and no one has ever documented significant decline, this is clearly an area of unregulated mariculture.

Needless to say, the Exxon Valdez disaster and more recent spills closer to home cast a long, dark shadow over all coastal areas vulnerable to such incidents. And which are not?

PLACES TO VISIT

Wherever there is access to coastal rocks and off-shore islands from Boston north, this community is well represented—for example, in the towns of Nahant, Marblehead, Gloucester, and Rockport. See *The Massachusetts Coast Guide* listed below. On the South Shore, Cape, and Islands, examples of the rocky intertidal community are scarce or absent.

Halibut Point State Park/Reservation, Rockport. Massachusetts Department of Environmental Management (DEM) and The Trustees of Reservations (TTOR).

World's End Reservation, Hingham. TTOR.

Peggoty Beach, Scituate. Town of Scituate.

Fort Phoenix State Reservation, Fairhaven. DEM.

FURTHER READING

At the Sea's Edge, by William T. Fox. New York: Prentice Hall, 1983.

Beachcomber's Guide to the North Atlantic Seashore. Lincoln, Mass.: Massachusetts Audubon Society, 1992. Pocket laminate.

The Edge of the Sea, by Rachael Carson. Boston: Houghton Mifflin, 1979.

Field Book of Seashore Life, by R. W. Miner. New York: G. P. Putnam's Sons, 1950.

A Field Guide to the Atlantic Seashore, by Kenneth L. Gosner. Boston: Houghton Mifflin, 1979.

Illustrated Key to the Seaweeds of New England, by Martine Villalard-Bohnsack. Kingston, R.I.: The Rhode Island Natural History Survey, 1995.

The Massachusetts Coast Guide, Access to Public Open Spaces Along the Shoreline, vol. 1, Greater Boston Harbor and the North Shore. Boston: Commonwealth of Massachusetts, 1995.

The Sea Is All About Us, by Sarah Fraser Robbins and Clarice Yentsch. Salem, Mass.: Peabody Museum, 1973.

A Sierra Club Naturalist's Guide to the North Atlantic Coast, by Michael and Deborah Berrill. San Francisco: Sierra Club Books, 1981.

Barrier Beach and Dunes

Sanderling

The sand here has a life of its own, even if it is only a life borrowed from the wind.

HENRY BESTON
The Outermost House

The sand that softens much of the Massachusetts coastline and that makes up most of the surface land mass of our southeastern coastal plain, Cape Cod, and the islands to the south was delivered by the Wisconsin ice sheet, the most recent of the great glaciers to cover New England. As the ice began to melt about 10,000 years ago, outwash streams sorted out enormous quantities of rock particles of all sizes, dropping the fine sands as the streams slowed down over the coastal plain and flowed into the sea. Longshore currents running north to south along the coast further dispersed these sands, depositing them on mounds of glacial till that form the skeleton of Cape Cod and our other "outer lands" and creating the long sand spits technically known as barrier beaches.

Our beach sands are mainly tiny fragments of granite minerals ground and sorted by abrasion, wind, and water. The semitransparent grains that can be distinguished under a hand lens are quartz. The milkier white and pink grains are feldspar. The purplish bits that make occasional lavender swaths on the upper beach are garnet. And black sand is an iron ore called magnetite that can be picked up with a pocket magnet. (Barrier beaches can also be composed of cobbles or pebbles, called shingle, or of shell or coral fragments.)

When a sand spit extends across an indentation in the mainland it creates a sheltered bay or lagoon. Typically this area fills with fine silt and sand washed from the land, creating tidal mud flats. Cordgrass (*Spartina alterniflora*) eventually takes root in the mud, signaling the birth of a salt marsh. The sandspit grows as wave action dumps sand from offshore. When the crest of the spit has risen to a certain height, the sand on top dries out and becomes subject to the almost constant breath of coastal winds. When the windblown sand meets a piece of debris—a gull carcass, an old sneaker—it accumulates into a small mound, which gathers ever more sand the higher it gets; this is the beginning of a sand dune. Once the dune has risen above the highest tide line the first pioneer plants take root. The most important of these is dune-grass, or "marram," whose generic name, *Ammophila*, is Greek for "sand lover." As more sand builds around it the grass simply extends its root system, keeping the leaves and flower heads on top. This root system, usually showing prominently on the steep leeward face of the dunes, is the main anchor of these shifting coastal hills. The

Black-bellied plovers

highest sand dunes in the world rise to about 1,500 feet in the coastal desert of Namibia, along the southwestern shore of Africa. In Massachusetts, the distinction goes to Mount Ararat in Truro, which crests at about 100 feet.

The characteristic shape of a dune reflects the direction of the prevailing wind: Sand builds up in a gentle slope on the windward side and then blows off the crest and falls onto the steep slope of the leeward side. As a dune system ages, excess sands often create one or more secondary dunes with sheltered troughs in between. The inland-facing dune is called the back dune; because of its relatively sheltered position, it often supports more robust and diverse vegetation. The seaward edge of the dune is called the toe of the dune.

Barrier beaches and dunes are ever-changing land forms. They are often called fragile, but "dynamic," "unpredictable," and "inexorable" may describe them better. The specialized organisms that live in these shifting sands are adapted not only to desertlike extremes of dryness and temperature, but also to drastic alteration in the landscape itself as winter seas devour sections of dune and winds bury thriving plant communities under drifts of sand.

Several distinct though interrelated natural communities are identifiable within the larger beach-dune ecosystem: The *intertidal zone* runs from the mean high tide line to mean low tide line; it supports no noticeable plant life, but contains numerous invertebrates living in the sand or foraging on top of it at high tide. At low tide many species of shorebirds feed actively in this wave-washed strip. The *wrack zone*, between the mean high tide line and the highest spring tide line at the toe of the dune, contains a few hardy plants, and its upper edge is breeding habitat for several rare birds and insects. The wrack itself, composed mainly of decomposing seaweeds, is the habitat of many small invertebrates such as beach fleas (amphipods) and flies, which in turn provide food for shorebirds and other predators.

The extremely dry *dunes* themselves support their own distinctive community of hardy shrubs and other plants and a few animals. *Interdune swales,* also called dune bogs or cranberry marshes, occupy the lowest troughs between two dune crests. Those lying below the water table may hold standing water year-round. Others fill up in winter and spring, but dry up by late summer. They tend to have a slightly enchanted quality: they are lush oases hidden among mountains of barren sand. Because of the poor, acidic soils many of the plants in interdune swales are also typical of bogs, with cranberry normally being the dominant species. On Cape Cod swales may also contain a few coastal plain pond rarities, such as Plymouth gentian and thread-leaved sundew. Depending on the depth of the soil and the water regime, some swales develop into shrub swamps, but the most interesting ones are colonized by a variety of grasses, sedges, and wildflowers that emerge through the cranberry mat.

Salt ponds may or may not be associated with barrier beach systems. In some cases they originated as lagoons between barrier beaches and salt marshes that became closed off by shifting coastal sands. In other cases they began as outwash streams that were sealed off following the glacial retreat. The long ponds that score the south shores of Nantucket and Martha's Vineyard are textbook examples of the latter. The uniqueness of salt ponds resides in their ambivalent position between saltwater and fresh water and the variations in salinity that occur from season to season or year to year or between the inland and seaward ends of the same pond. The variations depend on how much fresh water drains into the ponds while they are closed to the sea and, conversely, how often they get saltwater infusions when storm waves either wash over into them or open them temporarily to the sea. At their saltiest, salt ponds are colonized by many of the plants and animals that inhabit salt marshes and barrier beaches; at their

sweetest they may closely resemble freshwater marshes and ponds. As with coastal plain ponds (p. 68), the most significant natural feature of these ponds is their shoreline plant community, which becomes apparent when water levels are low. Atlantic mudwort (a lilliputian rarity, absent from most field guides), dwarf spikesedge, seaside crowfoot, false pimpernel, seaside flatsedge, little waterwort, and water pygmyweed are key indicator species.

The best-developed, most diverse examples of *dune maritime forest*, or "sunken forests," growing where roots have access to unsalty groundwater, contain surprisingly rich stands of tree species, including oaks, American beech, American holly, sassafras, and maples. These are typically filled with a tangle of vines, such as catbrier and wild grape, and a sparse shrub layer. Another form of maritime forest is dominated by dense stands of red cedar (juniper), sometimes accompanied by post and scrub oaks, black cherry, bayberry, and beach plum.

The wrack and intertidal zones and dune washout areas are obligatory nesting and feeding habitat for the globally rare piping plover and northeastern beach tiger beetle and the locally rare least tern, as well as the American oystercatcher. Roseate, Arctic, and common terns and black skimmers frequently nest in and around dunes and feed young in the intertidal zone. Barrier beaches are also important feeding and roosting areas for migratory shorebirds (sandpipers and plovers). Their chief food is invertebrates, both marine forms that live in the intertidal zone and species that breed or feed in decaying seaweed in the wrack line. They spend the hours during high tide roosting at the base of the dunes or in over-wash areas as well as in hummocks in nearby salt marsh. These beach habitats are crucial stopovers along migration routes that take these birds from arctic and subarctic breeding grounds to wintering grounds in Central and South America. The sanderling is the quintessential beach sandpiper, feeding habitually and hyperactively at the edge of the surf. Interdune swales sometimes contain concentrations of rare plant species. Maritime forests attract scores of migratory songbirds in May and again from August through October. Tree swallows migrating to neotropical wintering grounds swarm by the tens of thousands over coastal dunes in August and September, feeding on insects and bayberries and roosting in low shrubs and on the ground.

The pleasures of ocean beaches in summer draw millions of people to the Massachusetts shore annually, not for the most part to explore nature but simply to enjoy the salt air, the bracing water, the soft sand, and the cries of seabirds. It seems to be the natural habitat that humans like best. The dunes with their hidden forests and wetlands are known to many fewer people but contain their own strange magic found nowhere else.

Whimbrels

INDICATOR SPECIES

❧ **TREES:** Found almost exclusively in sheltered depressions, sometimes forming "sunken forests" in back dunes. Pitch pine (D, S), white pine (MF), red cedar (MF), quaking aspen (MF, S), black willow (MF), gray birch (S), scrub oak (D), black oak (MF, D), white oak (MF), red oak (D), American beech (MF), American holly (MF), basswood (MF), hackberry (D, MF), sassafras (MF, S), tupelo (MF), red maple (MF, S), witch hazel (MF), flowering dogwood (MF), pin cherry (D), black cherry (D, MF), red ash (MF). ❧ **SHRUBS AND VINES:** Found mainly in sheltered areas and wet depressions in the back dunes and as a shrub layer in maritime forests. Long-beaked willow (D), meadow willow (S), shining willow (S, D), common greenbrier or catbrier (D, MF), sweet gale (S), smooth alder (S), speckled alder (S), common barberry (D, X), wild currant (D, S), common (or thicket) shadbush (D, MF, S), smooth shadbush (D, MF), running shadbush (D, S), beach plum (D, MF), red chokeberry (S), salt spray rose (D, X), swamp dewberry (S, MF), wild raspberry (MF, S), various blackberry varieties (D, S), meadow sweet (S), steeplebush (S), Virginia creeper (MF, S), fox grape (MF), bayberry (D), swamp azalea (MF), bearberry (D), black huckleberry (D, MF), sheep laurel (S), maleberry (S, D), early lowbush blueberry (D), highbush blueberry (S, MF), large cranberry (S; dominant), winterberry (MF, S), winged sumac, poison ivy (D, MF), European buckthorn (S, X), Morrow honeysuckle (D, X), wild raisin (S), northern arrowwood (S). ❧ **FERNS AND CLUBMOSSES:** Southern bog clubmoss (S), cinnamon fern (S, MF), sensitive fern (MF, S), marsh fern (S). ❧ **GRASSES, SEDGES, AND RUSHES:** Dunegrass (D), seabeach quackgrass (W, D), little bluestem (D), seabeach needlegrass (D), poverty grass (D), hair fescue (D), nodding fescue (MF), depauperate panic-grass (D), fascicled panic-grass (D), saltmarsh switchgrass (D), six-weeks fescue (D), sea lyme-grass (D, SE), beachgrass (D, X), purple sandgrass (D), sand sedge (D), seaside flatsedge (S, SP), dune flatsedge (D), Small's spikesedge (S), dwarf spikesedge (SP), woolgrass or common bullsedge (S), white beaksedge (S), brown beaksedge (S), Pennsylvania sedge (D), dune sedge (D, X), seabeach sedge (D), broom-sedge (D), bristly sedge (S), beaded broom-sedge, saltmarsh straw-sedge (S), sharp-fruited rush (S), jointed rush (S), brackish rush (S), Greene's rush (S). The abundant and widespread toad, marsh, and soft rushes also occur occasionally in interdune swales. ❧ **WILDFLOWERS:** Yellow-eyed grasses (S), Canada mayflower (MF), star-flowered Solomon's-seal (D), grass pink (orchid; S), rose pogonia (orchid; S), seaside crowfoot (SP), nodding ladies'-tresses (orchid; S), seaside spurge (W), sand jointweed (D), Arethusa (orchid; S, ST), ragged fringed orchis (S), several species of pinweeds (W, D), common saltwort (W), seabeach orach (W), Drummond's arabis (D), sea rocket (W), wild radish (W, X), seabeach dock (W), seablite (W), grove sandwort (MF),

---

KEY

D=Dunes	SP=Salt Pond
FE=Federally Endangered	SE=State Endangered
FT=Federally Threatened	ST=State Threatened
IZ=Intertidal Zone	SSC=State Special Concern
MF=Maritime Forest	W=Wrack Zone
S=Swales	X=Exotic

FUNGI: Sandy laccaria, rough-stemmed boletus, hygroscopic earthstar. ∾ MOLLUSKS: Common species inhabiting the intertidal zone are northern moon snail, lobed moon snail, Stimson's whelk, chambered whelk, knobbed whelk, boat shell, qua-hog, surf clam, mahogany clam, common razor clam, ribbed pod, false angel wing, prickly jingle shell. The shells of deepwater-, mudflat-, or rocky bottom-inhabiting mollusks may also be found washed up on barrier beaches. ∾ DRAGONFLIES: The odonate (dragonfly and damselfly) fauna of dune swales is little studied. Several species of migratory dragonflies are occasionally abundant, flying along the dune line or swarming in sheltered hollows, including green darner, swamp darner, wandering glider (the only dragonfly that occurs worldwide), spot-winged glider, black-mantled glider. ∾ GRASSHOPPERS: Seaside locust, sand locust. ∾ BUTTERFLIES: There are no true barrier-beach butterflies, though the food plants of a number of common species occur in large dune systems. However, sea beaches and adjacent wild-flower areas are promising places to look for migratory species such as monarch, buckeye, painted lady, red admiral, snout butterfly (vagrant), cloudless sulphur, and little yellow (vagrant). ∾ TIGER BEETLES: Three species of these predaceous beetles inhabit barrier beaches in Massachusetts,

Quahog

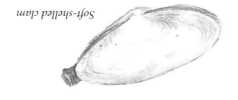

Soft-shelled clam

seabeach sandwort (W), water pygmyweed (SP), spatulate-leaved sundew (S), thread-leaved sun-dew (S), waxy meadow-rue (MF), wild strawberry (S), beach pea (W, D), swamp rose-mallow (SP), wild geranium (MF), northern St. John's-wort (S), orange grass (SP), little waterwort (SP), false heather (D), golden heather (D), maritime pin-weed (D), enchanter's nightshade (MF), saltpond pennywort (SP), small-flowered evening primrose (D), Plymouth gentian (S, SSC), bristly sarspar-

Northern moon snail

illa (D), wild sarsaparilla (D, MF), hemp-nettle (D, X), northern bugleweed (D), American german-der or wood sage (D), false pimpernel (SP), At-lantic mudwort (SP), cleavers (S), horse gentian (MF), dusty miller (D, X), tall wormwood (D), saltmarsh fleabane (SP), many-flowered aster (D), stiff aster (D), white wood aster (MF), wild lettuce (D), tall blue lettuce (D), seaside goldenrod (W, D), tansy (D, X), beach clotbur (W). ∾ LICHENS: A number of fruticose species are common in sheltered areas of bare sand, especially Iceland moss, reindeer moss, British soldiers. Old man's beard hangs in green wisps on trees and shrubs. ∾

Horse mussel

including the Federally Threatened/State Endangered northeastern beach tiger beetle. All of them live in holes in the upper beach as larvae and range down into the intertidal zone as adults. ∾ TRUE FLIES: Seaweed flies (W), great reddish robberfly. ∾ SPIDERS: Pike's or sand-dune wolf spider (D). ∾ CRUSTACEANS: Green crab (IZ), lady crab (IZ), flat-clawed hermit crab (IZ), long-clawed hermit crab (IZ), sand shrimp (IZ), beach fleas (W). ∾ AMPHIBIANS: Eastern spadefoot (toad; ST) aestivates in sand depressions in large dune areas and emerges to mate and breed in temporary pools after heavy rains in summer, an irregular phenomenon rarely witnessed. ∾ REPTILES: Hognose snake, smooth green snake (locally abundant in the Elizabeth Islands and on Martha's Vineyard). ∾ BIRDS: Piping plover (FT), American oystercatcher, roseate tern (FE), least tern (SSC), Arctic tern (SSC), common tern (SSC), black skimmer. All these breed on beaches and in dunes. Migratory shorebirds regularly found on Massachusetts beaches are black-bellied plover, semipalmated plover, ruddy turnstone, red knot, sanderling, semipalmated sandpiper, western sandpiper, least sandpiper, white-rumped sandpiper, and dunlin. Raptors, especially northern harrier (ST), merlin, American kestrel, peregrine falcon (FE), snowy owl (winter; SE), and short-eared owl (SE) favor dunes as hunting grounds. Tree swallows swarm by the tens of thousands over coastal dunes on migration in August and September; Savannah and vesper sparrows nest in open dune hollows. During periodic "irruptions" from the north, red crossbills are often found in mature coastal stands of pitch and Japanese black pine. ❧

MAMMALS: Gray seals pup on sandbars off Muskeget Island, northwest of Nantucket, and Monomoy Island in January and February and loaf on the beaches of Nantucket Sound during the remainder of the year. The beach, or Muskeget, vole, whether a separate species or distinctive subspecies of the meadow vole, lives on the island for which it is named. Red, hoary, and silver-haired bats migrate south along the coast in fall and often use barrier-beach thickets for roosting. Common mammals of large dune systems include meadow vole, meadow jumping mouse, white-footed mouse, eastern chipmunk, red squirrel (pines), white-tailed deer, red fox, long-tailed weasel, and eastern cottontail.

DISTRIBUTION

Barrier beaches and dunes occur worldwide wherever fine sediments, a longshore current, and appropriate coastal topography coexist. Along the East Coast of North America, these ecosystems form an almost continuous chain from southern Maine to Florida. In Massachusetts the chain is interrupted here and there by rocky headlands such as Cape Ann.

Tiger beetle

EXTENT IN MASSACHUSETTS

There are 684 barrier beaches in Massachusetts with a total area of 18,888 acres. Over 75 percent of these are small, under 10 acres, but the largest exceed 25 miles in length and comprise hundreds of acres. The top five, in order of acreage, are Monomoy, Chatham; Sandy Neck, Barnstable; Nauset Beach, Eastham and Orleans; Coatue, Nantucket; and Plum Island, Essex County.

CONSERVATION STATUS

Many of the barrier beaches of the Commonwealth are protected, with access controlled by government agencies or private conservation organizations. Heavy visitation by people is highly seasonal and tends to be restricted to portions of larger beaches near parking lots. By Massachusetts law all citizens may use the area between the mean high and mean low tide lines for purposes of "fishing, fowling, and navigating." However, unlike most other states, Massachusetts permits private ownership of land to the low tide line, and there is no guarantee of *access* over private land to get to the intertidal zone for the above-stated purposes. In many places stone revetments have been constructed in an effort to hold sand on certain favored beaches. This benighted practice disrupts the natural movement of coastal sediment, inevitably starving some beaches at the expense of others.

A major threat to wildlife is overuse of and damage to the wrack zone, especially where motor vehicles are still permitted to drive on the beach. Vehicles compact the soil, which inhibits the growth of plants and soil invertebrates such as tiger beetle larvae. They also destroy the wrack line

Piping plover

and, most significant, create tire ruts. These trap piping plover and least tern chicks, which may then be crushed when the next vehicle passes. Heavy foot traffic in the dunes kills beachgrass, thereby weakening the dune and altering the habitat.

Despite the contempt expressed in the Bible for the "foolish man, which built his house upon the sand" (Matthew 7:21), beach houses remain fashionable. Often the price for colonization of such waterfront real estate is contamination of groundwater and nearby coastal areas as a consequence of failed septic systems.

PLACES TO VISIT

Parker River National Wildlife Refuge (Plum Island), Essex County; access via Newburyport. U.S. Fish and Wildlife Service.

Crane Wildlife Refuge, Ipswich. The Trustees of Reservations (TTOR).

Plymouth Beach. Town of Plymouth.

Horseneck Beach State Reservation. Westport. Massachusetts Department of Environmental Management.

Sandy Neck, Barnstable. Town of Barnstable.

Cape Cod National Seashore, the outer Cape Cod beaches from Provincetown to Chatham; see especially the Provincelands dunes, near Provincetown, and Nauset barrier beaches in Eastham and Orleans. National Park Service.

Great Point, Nantucket. TTOR.

Cape Poge, Martha's Vineyard. TTOR.

FURTHER READING

At the Sea's Edge, by William T. Fox. New York: Prentice-Hall, 1983.

Barrier Island Handbook, by S. P. Leatherman. Amherst, Mass.: University of Massachusetts Press, 1979.

A Beachcomber's Botany, by Loren C. Petry and Marcia G. Norman. Chatham, Mass.: Chatham Conservation Foundation, 1963.

Beachcomber's Guide to the North Atlantic Seashore. Lincoln, Mass.: Massachusetts Audubon Society, 1992. Pocket laminate.

The Edge of the Sea, by Rachael Carson. Boston: Houghton Mifflin, 1979.

A Field Guide to the Atlantic Seashore, by Kenneth L. Gosner. Boston: Houghton Mifflin, 1979.

"Forgotten Habitats of Southern New England: Interdune Swales, Maritime Forests and Salt Ponds" by Bruce A. Sorrie. Wildflower Notes 6 (1991), no. 1., pp. 2–9.

Geologic Guide to the Cape Cod National Seashore, edited by S. P. Leatherman. Amherst, Mass.: National Park Service, 1979.

A Geologist's View of Cape Cod, by A. N. Strahler. Garden City, N.Y.: Doubleday, 1966.

The Great Beach, by John Hay. New York: W. W. Norton, 1980.

Life in the Shifting Dunes, by Lawrence B. White, Jr. Boston: Museum of Science, 1960.

The Outermost House, by Henry Beston.

Seashells of North America, by R. Tucker Abbott. New York: Golden Press, 1968.

A Sierra Club Naturalist's Guide to the North Atlantic Coast, by Michael and Deborah Berrill. San Francisco: Sierra Club Books, 1981.

The Wild Edge: Life and Lore of the Great Atlantic Beaches, by P. Kopper. New York: Times Books, 1979.

The Winter Beach, by Charles Ogburn. New York: William Morrow, 1966.

Running shadbush

Salt Marsh

Salt marshes, Plum Island

Where gaping mussels, left upon the mud,

Slope their slow passage to the fall'n flood:

Here…he'd lie down and trace

How side-long crabs had crawled their crooked race;

Or sadly listen to the tuneless cry

Of fishing-gull or clanging golden-eye;

What time the sea-birds to the marsh would come,

And the loud bittern, from the bull-rush home,

Gave from the salt-ditch-side the bellowing boom

GEORGE CRABBE
The Borough

S alt marshes are essentially coastal grasslands that grow between the tides. They typically form in lagoons behind barrier beaches and in other sheltered localities such as bays near the mouths of estuaries. In most salt marshes, several well-marked zones, defined by relative amounts of flooding and salinity, can be recognized. The great expanse of relatively short grass that covers most of the salt-marsh landscape, whipped by wind and water into distinctive tufts, is called high marsh and is dominated by perennial saltmeadow cordgrass, or salt hay. It is flooded only during the highest monthly (spring) tides. At the edges of tidal creeks, the fringes of salt-marsh pools, or the shores of mud flats, a usually narrower area of low marsh prevails and is readily identified by the presence of the much taller saltwater cordgrass. This zone gets flooded twice daily, at each high tide. Along the upland edge of the high marsh, where the soil is driest but is still somewhat saline, there is often a border of the graceful common reed, freshwater cordgrass, saltmarsh elder, and/or narrowleaf cattail, often mixed with a variety of salt-tolerant wildflowers such as sea lavender or seaside gerar-

dia. Salt marshes are filled and drained by meandering tidal creeks and pocked with shallow pools, or *pannes*, both of which are crucial elements in the life of this rich ecosystem.

Salt marshes are formed when pioneer seeds of saltwater cordgrass take root in tidal sediments or when rafts of peat containing living plants are plucked out by winter ice and deposited where rooting conditions are favorable. An acre of this grass can produce ten tons of plant material a year, and the accretion of dead plant material that rapidly accumulates eventually forms high peat banks that may rise above the mean high water mark and provide the necessary growing conditions for the high marsh–dominant saltmeadow grass.

The peaceful grandeur of a large estuarine salt marsh masks the orgy of biological activity taking place in and around it. It is an eco-truism that this community is one of the most biologically productive on the planet, crucial to the well-being of an astonishing range of organisms, from lowly bacteria and marine fishes to migratory birds and the great hominid ape. The keystone phenomenon of this impressive system is the prodigious

Greater yellowlegs

amount of detritus produced by cordgrasses. This natural compost is the mainstay of hosts of worms, crustaceans, mollusks, insect larvae, young fish, and other small organisms, many of which will themselves make a meal for a larger predator. The most nutritious component of the detritus is not the plant material itself, but rather the bacteria and fungi that decompose it. Once these decomposing organisms have been gleaned from fresh detritus in the guts of detritivores such as fiddler crabs, the undigested remainder is excreted, attacked once again by bacteria and fungi, and then reconsumed by other organisms. The energy of this humble grass thus fuels the entire system, from bacteria and fungi to crab larvae to small fish to larger fish and through an osprey and an infinite number of analogous sequences. Since a significant fraction of commercial shell- and fin-fish species in the Northeast—soft-shelled clam, winter flounder, striped bass, and bluefish, to name a few—depend on salt marshes during part of their lives, we too are the beneficiaries in no small measure of the mighty salt-marsh grasses.

The force that drives the detritus system, helping to build the structure of the marsh and moving food around in it, is the ocean tides.

Salt marshes are an inextricable component of the sensual intoxication to be found in the quieter havens of the New England coast. These grasslands and the great haystacks placed on closely driven pilings, or staddles, above the spring tide line were favorite subjects of the painters of the New England Luminist school such as Martin Johnson Heade. His famous painting "Sunrise on the Marshes" (ca. 1865–1870), now on loan to the White House, conveys the splendor of these fertile expanses. Anyone who lives near these sea-drowned prairies has internalized a host of indelible impressions: canoeing into the silent heart of the marsh via a meandering creek; squadrons of tree swallows hawking for mosquitoes in August; a snowy owl perched on a hay staddle in January; an unmistakable sweet tang in the nostrils; the dawn song of the seaside sparrow; the surprisingly painful bite of a greenhead fly; catching mummichogs with a dip net; watching a merlin plunge into a mixed flock of shorebirds; the mechanical jousting of fiddler crabs; the October scarlet of samphire....

INDICATOR SPECIES

Most of the plants listed are "obligate species," that is, they occur only in salt marshes. Most species described live in the high marsh.

~ SHRUBS: Marsh elder, groundsel tree. ~ GRASSES, SEDGES, AND RUSHES: Saltmeadow grass or salt hay (high marsh dominant), saltwater cordgrass (low marsh dominant), freshwater cordgrass (upland edges), common reed or Phragmites (upland edges), saltmarsh spikegrass, saltmarsh alkaligrass, saltmarsh wild-rye, saltmarsh switchgrass, saltmarsh rush or blackgrass, brackish rush, seaside toad-rush, saltmarsh flatsedge, seaside flatsedge, saline spikesedge, saltmarsh spikesedge, dwarf spikesedge (salt pond edges), saltmarsh threesquare, brackish bullsedge, seaside bullsedge, saltmarsh bullsedge, saltmarsh strawsedge, saltmarsh sedge, eastern saline-sedge. ~ WILDFLOWERS: Widgeon grass, saltmarsh arrowgrass, sea lavender, seaside plantain, saltmarsh sand-spurrey, sea milkwort, sea blite (3 species), marsh orach, samphire or common glasswort

KEY
FE=Federally Endangered
SE=State Endangered
ST=State Threatened
SSC=State Special Concern

(colonizes bare areas of salt pools), woody glass-wort (edges of salt pools), dwarf saltwort, water hemp, marsh mallow, rose mallow, annual marsh pink, silverweed, seaside gerardia, saltmarsh amaranthus, saltmarsh fleabane, seaside goldenrod, annual or large saltmarsh aster, perennial salt-marsh aster. ❧ MOLLUSKS: Common coffee bean snail, common and rough periwinkles, eastern mud nassa, soft-shelled clam. ❧ DRAGONFLIES: Saltmarsh skimmer. ❧ GRASSHOPPERS: Dusky-faced meadow grasshopper, saltmarsh meadow grasshopper. ❧ TRUE FLIES: Saltmarsh mosquito, greenhead fly, chironomid midges, biting midges. ❧ BUTTERFLIES: Common wood nymph (caterpillar feeds on cordgrass), broad-winged skipper (Phragmites). ❧ SPIDERS: Emerton's and other dwarf spiders, species of sac spiders. ❧ CRUSTACEANS: Grass shrimp, sand shrimp, green crab, fiddler crabs (several species), the isopod *Philoscia viltata, orchestia* beach fleas (amphipods), horseshoe crab. ❧ FISHES: Sheepshead minnow, mummichog, spotfin killi-fish, rainwater killifish, inland silverside, Atlantic silverside, fourspine stickleback, threespine stickleback, blackspotted stickleback, ninespine stickle-back. In addition, a number of anadromous fishes (e.g., striped bass, American eel) or marine fishes (e.g., Atlantic tomcod, hogchoker) frequently occur in estuarine salt-marsh creeks. These waters are also important feeding grounds for juvenile fish of a number of commercial species such as striped bass and winter flounder. ❧ REPTILES: Diamond-backed terrapin (Cape Cod and south). ❧ BIRDS: Salt marshes are very rich in birdlife, though relatively few species are restricted to them. Here, important groups and a few species are highlighted. All of the wading birds (herons, bittern, ibises) that occur in Massachusetts use salt marshes at least occasionally. Great and snowy egrets are particularly conspicuous. Great blue herons resort to coastal marshes during migration or when inland wetlands freeze up, and remain until the salt marshes are clogged with ice. Spring tides on salt marshes provide excellent opportunities to see the normally reclusive American bittern (ST), especially during fall migration. Most of our native dabbling ducks also frequent salt-marsh pools; these are particularly favored by black ducks and Canada geese—which in turn are much favored by waterfowl hunters. Raptors are also attracted to salt marshes because of the abundance of prey such as voles, small birds, fish, and insects. These raptors include osprey (breeds in salt marshes), rough-legged hawk (winter), northern harrier (ST), peregrine falcon (FE), merlin, snowy owl (winter), and short-eared owl (SE). Willets are now nesting again in our salt marshes, having been "shot out" by the market gunners of the nineteenth century. In addition, most of the migrant shorebirds that visit Massachusetts in spring and fall feed in salt-marsh pools and on mudflats along marsh creeks; the plaintive whistles of the greater and lesser yellowlegs and whimbrel and the high peep of the least sandpiper are among the most conspicuous voices of the salt marsh. Common terns (SSC) regularly nest in isolated parts of the marsh, and these and least terns (SSC) frequently fish in salt-marsh pools. Gulls use the pools mainly for loafing. Eastern kingbirds, purple martins, and tree swallows seem to choose nest sites near salt marshes and feed constantly on the marsh's enormous populations of insects. True obligate salt-marsh birds in New England include clapper rail, willet, Wilson's phalarope, seaside sparrow, and saltmarsh sharp-tailed sparrow. The two sparrows build their nests in the upper reaches of saltwater cordgrass and regularly get flooded out during the highest tides; their relatively long, narrow bills and short tails are adaptations to perching and feeding in the marsh. ~

MAMMALS: Meadow vole.

DISTRIBUTION

The form of cordgrass-dominated coastal marshes we call salt marsh in Massachusetts is largely restricted to the Atlantic coast of eastern North America from the Gulf of St. Lawrence to Florida. It runs almost continuously from New Hampshire southward. Examples of salt marsh exist in almost every Massachusetts coastal town, including the city of Boston. Extensive examples are Rumney Marsh, Saugus; North and South River estuaries, Scituate and Marshfield; Duxbury Marsh; Great and Little Sippewissett Marshes, Falmouth; the Great Marshes, Barnstable; and Nauset Marsh, Eastham and Orleans on outer Cape Cod.

EXTENT IN MASSACHUSETTS

According to Hankin et al. (1985), at present Massachusetts contains just over 48,000 acres of salt marsh. The largest continuous system, comprising about 20,000 acres, is the Essex Bay–Plum Island Sound expanse between Gloucester and Newburyport.

Saltmarsh sharp-tailed sparrow

Willet

CONSERVATION STATUS

A large proportion of Massachusetts salt marsh has been destroyed since European colonization. Much of what is now Boston, including Back Bay and the Fenway area, was once salt marsh that filled the Charles and Mystic River estuaries.

Perhaps the most unfortunate episode in the history of salt-marsh management was the largely successful effort to ditch and drain these wetlands, ostensibly for mosquito control, mainly during the Great Depression when an abundance of labor was available. Ditching alters normal drainage patterns and eliminates many of the salt-marsh pools which are breeding areas not only for mosquitoes,

but also for the mosquitoes' main predators such as small fish and predaceous insects. These pools are also prime feeding areas for many species of waterbirds and other wildlife.

In the period of rapid development following World War II, the Commonwealth lost approximately a third of its then existing salt marsh: 20,000 acres between 1945 and the mid-1970s.

Marsh degradation, especially near urban areas, has also occurred as a result of disruption of tidal flow, filling, pollution, and other forms of disturbance such as fire. Such deterioration is often signaled by the proliferation of common reed (Phragmites), which replaces the cordgrass, greatly reducing the natural biological diversity of the wetland.

Since the 1970s, destruction of salt marsh has slowed, thanks in part to very strict laws that allow virtually no alteration of these systems. Even so, recent estimates (Buchsbaum, ed., 1992), indicate an additional loss of 4 percent since 1978.

PLACES TO VISIT

Essex Marshes, numerous access points in Ipswich, Gloucester, Essex; marsh cruises run spring through fall from the Town of Essex. Various public and private owners.

Parker River National Wildlife Refuge (Plum Island), Essex County; road access from Newburyport. U.S. Fish and Wildlife Service.

Belle Isle Marsh, East Boston. Metropolitan District Commission.

Sandy Neck, Barnstable. Towns of Sandwich and Barnstable.

Wellfleet Bay Wildlife Sanctuary, Wellfleet. Massachusetts Audubon Society.

Nauset Marsh, Cape Cod National Seashore, Eastham and Orleans. National Park Service.

FURTHER READING

Barrier Beaches, Salt Marshes and Tidal Flats: An Inventory of the Coastal Resources of the Commonwealth of Massachusetts, by A. Hankin, L. Constantine, and S. Bliven. South Dartmouth, Mass.: Lloyd Center for Environmental Studies, 1985.

The Ecology of New England High Salt Marshes: A Community Profile, by Scott Nixon. U.S. Fish and Wildlife Service, Community Profile, Washington, D.C.: U.S. Government Printing Office, 1982.

A Field Guide to Coastal Wetland Plants of the Northeastern United States, by Ralph W. Tiner, Jr. Amherst, Mass.: University of Massachusetts Press, 1987.

The House on Nauset Marsh, by Wyman Richardson. Old Greenwich, Conn.: Chatham Press, 1947; reprint 1972.

The Life and Death of the Salt Marsh, by John and Mildred Teal. Boston: Little, Brown, 1971.

Salt Marshes and Salt Deserts of the World, by V. J. Chapman. New York: Interscience Publications, 1960.

Turning the Tide: Toward a Livable Coast in Massachusetts, edited by R. Buchsbaum. Lincoln, Mass.: Massachusetts Audubon Society, 1992.

See also seashore natural history guides listed on p. 59 (BARRIER BEACH AND DUNES), many of which include sections on salt marsh.

This is the grass that grows wherever the land is and the water is…

WALT WHITMAN
"Song of Myself"

Coastal Plain Pond

Coastal plain pond, Myles Standish State Forest

Coastal plain ponds are typically shallow, often small bodies of water in kettlehole depressions in sandy soil that are fed by groundwater. The most significant ecological feature of these ponds is the floral community of the pond shore. The plants of this association, some of them globally rare, have evolved to thrive in a zone of radical and unpredictable fluctuations in water level. In dry years, when the ponds reach their lowest levels, the shores erupt in a profusion of wildflowers, while in the wettest years, when little or no beach is uncovered, the plants remain dormant underwater, until suitable conditions recur. Depending on the timing and duration of inundation and exposure, vastly different suites of species will flourish around the shore in different years. This is perhaps the most changeable plant community in the Commonwealth, putting one in mind of an indecisive gardener who changes the layout of her bedding plants every year. Some species sprout underwater in the spring from seed, roots, or perennial rosettes of leaves, while others will not germinate until the water recedes. In addition to the ups and downs of the water table, pond-shore plants must tolerate the general dryness, acidity, and lack of nutrients that are characteristic of the very well drained sandy soil and pond water.

No doubt many people have spent blissful summer hours along the shores of coastal plain ponds on the South Shore and Cape Cod, completely unaware of being in the presence of a globally significant ecosystem. Botanically speaking, the best ponds are the sheltered ones with little wave action to disturb the seed bank. These quiet hollows with their lenses of still water reflecting the surrounding pine-oak woods are a quintessential feature of the old, wild, coastal sandplains. In August and September, when the water has dropped, pink clouds of the globally rare but locally abundant Plymouth gentian envelop the shore. Other rarities make up in fascinating oddity what they lack in size and showiness. You will have to get down on all fours to inspect the thread-leaved sundew properly, but you may be rewarded by seeing this delicate carnivore digest a hapless damselfly before your eyes. Out over the pond's limpid surface watch for the great comet darner, arguably Massachusetts's most spectacular dragonfly. This scarlet and forest green aerialist is so hard to net that the early naturalists resorted to blasting it with a .410 and dust shot!

Plymouth gentian

∾ **PLANTS:** Characteristic species occur in zones between the high water mark before the upland shrubs begin and the edge of the water. The driest upper reaches may contain Maryland meadow beauty (SSC), thread-leaved sundew (SSC), spatulate-leaved sundew, and New England boneset (SE). Most of the typical coastal plain pond plants occur in the intermediate zone, including Plymouth gentian (SSC), golden hedge hyssop (or pert), pink tickseed (SSC), and slender-leaved goldenrod. Species emerging from the water or occurring in waterlogged soil are slender arrowhead (SSC), fibrous and two-flowered bladderworts, and often a variety of sedges, rushes, and pipeworts. ∾ **MOLLUSKS:** Ordinary spire snail, pupa-like spire snail, eastern lamp mussel, pointed sand-shell mussel. ∾ **INSECTS:** Comet darner dragonfly, New England and pine barrens bluets (damselflies, both SSC). Other aquatic insects may be endemic to these ponds, but the faunas have not been well studied. ∾ **VERTEBRATES:** The clear standout in this category is the Plymouth redbelly turtle (FE), known from only a few ponds in Plymouth County; an apparently successful captive breeding and reintroduction program is ongoing under the auspices of the Massachusetts Division of Fisheries and Wildlife. Otherwise, coastal plain ponds are usually inhabited by a wide range of common pond vertebrates (see LAKES AND PONDS), none of which are uniquely associated with this habitat.

KEY
FE=Federally Endangered
SE=State Endangered
ST=State Threatened
SSC=State Special Concern

DISTRIBUTION

Globally, coastal plain ponds are restricted to the Atlantic coastal plain of North America from Nova Scotia to Florida and Texas and to the Great Lakes sand barrens in southwestern Michigan and central Wisconsin. In New England the best examples of these ponds are in southern Plymouth County, on Cape Cod, on Nantucket, and in Rhode Island. Similar ponds with fewer pond-shore species occur in outwash sandplains in the southern Connecticut River Valley in Massachusetts.

EXTENT IN MASSACHUSETTS

According to the Massachusetts Natural Heritage and Endangered Species Program, there are probably 200 or more coastal plain ponds of high quality. Such ponds have the gradually sloping shorelines that lend themselves best to the development of the plant community and have not yet suffered degradation from development. Except for so-called depression meadows that support the plant community across the entire bottom of seasonally dry ponds, the habitat tends to be very narrow (6–30 feet) and highly variable in width, and the total acreage or linear dimension is therefore difficult to estimate.

CONSERVATION STATUS

Coastal plain ponds are among the Commonwealth's greatest botanical treasures, and contain many plants that reach the northernmost extremity of their range in Massachusetts. Sadly, a great many of these fragile ecosystems are being damaged or destroyed by human activities. Although this adaptable plant community depends on fluctuating water levels, its rare flora will eventually drown if high water levels are artificially maintained through many seasons; likewise it will shrivel away after too many dry years caused by water withdrawals. When the latter occurs, upland plants such as bayberry, highbush blueberry, and little bluestem grass, which are normally flooded out in wet years, establish themselves at the expense of the pond-shore specialties. Another threat is a rise in nutrient levels caused by leaking septic systems or too many waterfowl; this encourages the growth of algae and alien pond weeds, which use up the available oxygen and can choke out the rarer species. Perhaps the most widespread threat is overuse of the pond shore, especially by off-road vehicles. Even relatively infrequent trampling over the same areas can create compacted tracks where plants are unable to grow. Wholesale destruction from development is also a factor. The Mary Dunne Ponds in Hyannis, probably the most spectacular example of this rare community in the Northeast, were saved in the eleventh hour by the Massachusetts Division of Fisheries and Wildlife from being bulldozed for commercial development.

PLACES TO VISIT

If you visit any of these unique sites, please be sensitive to the concerns about trampling noted above. More than one sensitive plant community has been loved to death by well-meaning admirers.

Myles Standish State Forest (several ponds), Plymouth. Massachusetts Department of Environmental Management (DEM).

Massasoit State Park, East Taunton. DEM.

Nickerson State Park, Brewster. DEM.

Grassy Pond, Ashumet Holly and Wildlife Sanctuary, Falmouth. Massachusetts Audubon Society (MAS).

Sesachacha Heathland Wildlife Sanctuary, Nantucket. MAS.

Most towns on Cape Cod have swimming beaches on coastal plain ponds with the typical pond-shore flora, for example Flax Pond in Bourne, Hathaway Pond in Barnstable, and Dennis Pond in Yarmouth.

FURTHER READING

"A classification scheme for coastal plain pond-shore and related vegetation from Maine to Virginia," by L. A. Sneddon and M. G. Anderson. Boston: The Nature Conservatory, Eastern Regional Office, 1994.

Atlantic White Cedar Swamp

Many decades ago, [there were] vast areas of cedar...swamps where the trees stood so thick that semi-darkness prevailed beneath the canopy of their matted crowns and sphagnum moss grew luxuriantly over their roots. For untold centuries the moss leaves and other vegetation died and partially rotted, forming a heavy layer of peat over the soil. The agricultural possibilities of these peat swamps did not long go unrecognized by tillers of the soil and as rapidly as drainage and clearing could be done, they were converted to agricultural use. The inexcusable part of this conversion process was that fire, started on the clearing areas, was permitted to spread to the uncleared areas with the result that most of the cedar forests and the rich peat on which they grew have been destroyed.

WILLIAM S. TABER,
DELAWARE'S FIRST STATE FORESTER

*T*he classic form of this community is a dense, nearly pure stand of Atlantic white cedar growing in highly acid (pH 3.1–5.5) organic soils where few other New England tree species can thrive. These conditions and many of the plant and animal species associated with white cedar swamps are similar to those in bogs.

Like bogs, white cedar stands often occur in concentric wetland formations in kettleholes: swamp loosestrife emerging from open water around the perimeter, then a sphagnum mat (often populated with typical bog plants such as pitcher plant and sundews), then a shrub zone dominated by a variety of heath species and hollies followed by a stand of white cedar. In more open cedar stands these shrubs frequently occur as an understory, and other wetland tree species, especially red maple, may be mixed in. Mature cedars can reach a height of 85 feet and a diameter of 1.5 feet, but most such trees were put to the ax long ago in New England; in most stands today trees half this size or less are the norm.

Once familiar with the "field marks" of white cedar swamps—the blackish, somewhat forbidding mass of trees with a brassy greenish cast in sunlight—one will readily recognize them huddled in wet depressions as one drives across the southeastern coastal plain of southern New England. The somber beauty of this habitat is best experienced at a respectful distance. The interior of an Atlantic white cedar swamp is a kind of aquatic iron maiden experience: dark and cold, one's lower extremities sunk in malodorous icy wetness, all parts of the body pierced by sharp twigs. Only the masochistic botanist or lepidopterist in search of rarities finds contentment in this dank, disorienting world.

Opposite: Atlantic white cedar swamp, Wellfleet

INDICATOR SPECIES

∾ **TREES:** Atlantic white cedar (O; dominant), eastern hemlock, red maple, tupelo, yellow birch. Black spruce and tamarack in inland and more northern localities. ∾ **SHRUBS:** Common winterberry, smooth winterberry, inkberry, mountain holly, sweet pepperbush, swamp azalea, swamp highbush blueberry, swamp sweetbells, dangleberry, maleberry, large cranberry, poison sumac, dwarf mistletoe (a woody parasite on cedar and spruce; SE). ∾ **FERNS:** Virginia chain fern, netted chain fern, Massachusetts fern, royal fern. ∾ **WILDFLOWERS:** Heartleaf twayblade (O, SE; an orchid currently known from only one Massachusetts locality), goldthread, northern white violet, and other acid-tolerant species. ∾ **MOSSES AND LICHENS:** The herb layer is typically dominated by sphagnum mosses. Old man's beard lichen (*Usnea* species). See indicator list under BOG for other plant species that may occur in cedar swamps. ∾ **BUTTERFLIES:** Hessel's hairstreak (O, SSC). The underwings of this small butterfly are spectacularly patterned in red, green, and silver. Unfortunately, few people get to appreciate this splendid insect since the adults spend most of their lives in the tops of cedars; the caterpillar feeds exclusively on needles of Atlantic white cedar. ∾ **DRAGONFLIES, OTHER INVERTEBRATES:** See under BOG. ∾ **BIRDS:** Saw-whet owl, veery, northern waterthrush, Canada warbler. The northern parula (ST) uses old man's beard lichen to make its nest. Both lichen and bird appear to be declining in Massachusetts, but are currently maintaining strongholds in Atlantic white cedar swamps. ∾ **MAMMALS:** Red-backed vole, bobcat. Snowshoe hares have been introduced and have become successfully established in several Massachusetts cedar swamps.

Atlantic white cedar

DISTRIBUTION

The Atlantic white cedar is restricted to eastern North America, mainly to Atlantic coastal plain from central Maine (Knox County) to North Carolina with scattered stands in South Carolina, Georgia, and north-central Florida. A distinct variety or perhaps a separate species of Atlantic white cedar occurs along the Gulf Coast in Mississippi, Alabama, and Florida.

In Massachusetts, Atlantic white cedar swamps are concentrated mainly throughout the sandy coastal plain south and southeast of Boston, occurring in almost every town in this region. However, they also occur sporadically in the rockier terrain of Essex County and in disjunct populations in Worcester County and the Connecticut River Valley.

KEY
O=Obligate
SE=State Endangered
SSC=State Special Concern

EXTENT IN MASSACHUSETTS

According to Motzkin (1991), there are about 150 distinct Atlantic white cedar swamps in Massachusetts with a cedar component greater than 5 percent. There are 91 swamps larger than 3 acres that together total about 7,000 acres. Sixty of these sites each comprise more than 10 acres with a cedar component greater than 25 percent.

CONSERVATION STATUS

The durability of cedar stumps and the existence of an extensive economic record leave no doubt that the scattering of even-aged stands of second-growth Atlantic white cedar currently remaining in New England is a pitiful vestige of the magnificent forests that once occurred nearly continuously in suitable coastal areas. So valuable was this durable timber resource that during the colonial period ownership of cedar swamps was a measure of a landowner's wealth, and within early communities rights to the trees were apportioned to residents in deeded strips.

In addition to lumber harvesting and the mining of bog iron in swamps, draining and clearing of cedar swamps has occurred on a grand scale for the creation of commercial cranberry bogs. This practice alone has accounted for the destruction of more than 15,000 acres of wetlands in Massachusetts since the mid-eighteenth century. Many additional acres of cedar swamp have been cut and filled for residential and commercial development. As late as 1970 over 3,000 acres of prime cedar swamp on the South Shore were illegally "converted" to pasture and cornfield by a large dairy corporation.

In their work in the white cedar swamp at Marconi on Cape Cod, Patterson and Motzkin found no cedar reproduction within the stand. Perpetuation of the community appears to require fire to open it up. Because fires were more common in early colonial times, white cedar is more abundant now than it was prehistorically, but will presumably decline as a result of fire suppression.

Of the 91 sites noted above, 31 are at least partially on protected or public lands (Motzkin, 1991). All are at least nominally protected by the Massachusetts Wetlands Protection Act.

PLACES TO VISIT

Acushnet Cedar Swamp, New Bedford; a National Natural Landmark. Town of New Bedford.

Lakeville Cedar Swamp, Lakeville. Massachusetts Audubon Society.

Marconi Cedar Swamp, Cape Cod National Seashore, Wellfleet. National Park Service.

Cherry Street Bog, Ipswich River Wildlife Sanctuary, Wenham. Massachusetts Audubon Society.

Douglas State Forest, Douglas; good access via boardwalk. Massachusetts Department of Environmental Management.

FURTHER READING

"Atlantic White Cedar Swamps," by Glenn Motzkin. Amherst, Mass.: Massachusetts Agricultural Experiment Station Research Bulletin No. 731, 1991.

Atlantic White Cedar Wetlands, edited by Aimlee D. Laderman. Boulder, Colo., and London: Westview Press, 1987. (Including "Atlantic White Cedar in the Glaciated Northeast" by Laderman, Golet, Sorrie, and Woolsey and "The Status and Distribution of Atlantic White Cedar in Massachusetts" by Sorrie and Woolsey.)

Cedar of Acid Coastal Wetlands: Chamaecyparis thyoides *from Maine to Mississippi,* by Aimlee D. Laderman. New York: Van Nostrand Reinhold, 1988.

Sandplain Grassland

Upland sandpiper

*This island must look exactly like a prairie,
except that the view in clear weather is
bounded by the sea....*

HENRY DAVID THOREAU
describing Nantucket in *Journal 7*

The sands deposited by glacial rivers along the coast or by river sedimentation inland are our driest and poorest soils. The first colonizers of such harsh lands are grasses and a few specialized forbs. In very dry climates—the planet's natural prairies or steppes—more robust plant forms cannot survive and these hardy pioneers remain dominant. In New England, the climate is relatively humid, and all but the thinnest and wettest soils can support trees if left undisturbed. Grasslands occurring on uplands in Massachusetts, therefore, are all successional communities: If left alone they will eventually, slowly, grow up to shrubland and then to forest. For the most part, sandplain grasslands persisting in New England today have been strongly shaped by human activity, traditionally, fires. There are perhaps a few exceptions, where coastal salt spray and wind combine with dryness and poor soil to maintain this early successional stage naturally.

Many of the plant and animal species that occur in these grasslands, or their close relatives, also occur in the grasslands of the Midwest. Little bluestem, big bluestem, switchgrass, and Indian grass are major constituents of both the

Stiff aster and little bluestem grass

sandplains of Martha's Vineyard and the tall-grass prairies of Iowa and surrounding states.

A broad rolling expanse of prairie with waving grasses, patches of colorful wildflowers, birds singing on the wing beneath a blue, domelike sky, and the smell of rain in the air is one of the most pleasing landscapes on earth. Emotional responses aside, sandplain grasslands contain some of the greatest concentrations of rare species in New England, including the globally rare bushy rock-rose and sandplain gerardia, extraordinary birds such as the short-eared owl and upland sandpiper, and a rich diversity of butterflies, including the aptly named regal fritillary, now all but extinct in New England.

Sandplain grassland is closely associated with coastal heathland.

INDICATOR SPECIES

∾ SHRUBS: Sparsely scattered individuals of Nantucket shadbush and shrubs in the heath family. ∾ GRASSES AND SEDGES: Little bluestem, big bluestem, switchgrass, Indian grass, Pennsylvania sedge. ∾ WILDFLOWERS: Goat's rue, wild indigo, and bush clovers (all members of the pea family that bear nodules of nitrogen-fixing bacteria on their roots, which help them obtain sufficient nutrients from the sterile soil), bird's-foot violet, bluntleaf milkweed, sandplain blue-eyed grass (SSC), sandplain flax (SSC), sandplain gerardia (FE), bushy rockrose (SSC), northern blazing star (SSC), hyssop-leaved boneset, narrow-leaved and toothed white-topped asters. ∾ BUTTERFLIES:

KEY
FE=Federally Endangered
SE=State Endangered
ST=State Threatened
SSC=State Special Concern

regal fritillary (extirpated; caterpillar feeds on bird's-foot violet), wild indigo duskywing (wild indigo). Cobweb, dusted, Leonard's, and Indian skipper caterpillars feed on big and little bluestem grasses. ∾ BIRDS: Northern harrier (ST), upland sandpiper (SE), short-eared owl (SE), horned lark, vesper sparrow (ST), savannah sparrow, grasshopper sparrow (ST). ∾ MAMMALS: Meadow voles are abundant, providing the food for breeding and wintering raptors.

DISTRIBUTION
Strictly defined, New England sandplain grassland is an extremely rare and localized community, its global range essentially limited to coastal areas from southeastern Massachusetts to Long Island. By far the best examples occur on Nantucket, Tuckernuck, Martha's Vineyard, the Elizabeth Islands, and Cape Cod. The largest and best example of New England sandplain grassland in the world is the Miacomet Plains on Nantucket.

A similar but much less diverse sandplain grassland community occurs very locally on sand terraces along the Connecticut River.

EXTENT IN MASSACHUSETTS
There are fewer than a dozen existing occurrences of New England sandplain grassland, totaling approximately 6,000 to 8,000 acres.

CONSERVATION STATUS
Successional communities such as sandplain grassland have evolved over millions of years in response to natural disturbances and often contain concentrations of rare species. Before European settlement, wildfires set by lightning strikes or

Northern harrier

Native Americans were benign, natural events that burned hundreds or thousands of acres, creating sporadic openings of grasslands and shrublands in the dominant forest cover and thus ensuring the continual reemergence of the sandplain grassland community. Since we can no longer depend on spontaneous wildfires to create a mosaic of biologically diverse habitats in various stages of succession over extensive landscapes, we are left with the alternative of maintaining habitats such as sandplain grassland by retarding natural succession, using technologies such as prescribed burning. Many of the best sandplain grassland sites on Cape Cod, Nantucket, and Martha's Vineyard are now being intensely managed and monitored in this way, and it is gratifying to report that the many globally rare species of this community, whose existence is at stake, seem to be thriving on this peculiar kind of tender loving care.

PLACES TO VISIT

Sesachacha Heathland Wildlife Sanctuary, Nantucket. Massachusetts Audubon Society (MAS).
Miacomet Plains, Nantucket. Nantucket Conservation Foundation and the Nantucket Land Bank.
Katama Great Plains, Edgartown, Martha's Vineyard. The Nature Conservancy.
Wellfleet Bay Wildlife Sanctuary, Wellfleet. MAS.
Long Point Reservation, West Tisbury, Martha's Vineyard. The Trustees of Reservations.

FURTHER READING

Changing Landscapes: A Pictorial Field Guide to a Century of Change on Nantucket, by Peter W. Dunwiddie. Nantucket: Nantucket Historical Association, 1992.
Martha's Vineyard Landscapes: The Nature of Change, by Peter W. Dunwiddie. Vineyard Haven, Mass.: Vineyard Conservation Society, 1994.

Bird's-foot violet

Coastal Heathland

Precisely at this transitional point of its nightly roll into darkness the great and particular glory of Egdon waste began, and nobody could be said to understand the heath who had not been there at such a time. It could be best felt when it could not be clearly seen.

THOMAS HARDY
The Return of the Native

Coastal heathland is an open, virtually treeless community dominated by ericaceous shrubs—low-growing, woody members of the heath family. Coastal heathland is also known as maritime heathland, heath sand barrens, and moorland. This habitat prevails naturally along the New England coast, where dry, nutrient-poor soil, constant wind, and salt spray combine to prevent succession to taller, less hardy plant species. Elsewhere it is a *seral,* or successional, stage that will eventually give way to forest, especially pitch pine–scrub oak barrens. Because it is a very localized habitat in its natural state, many of the plants and animals that thrive only in this community are rare. Structurally, heathlands often form mosaics with other communities, clumps of heath shrubs alternating with swathes of grass and wildflowers, tufts of lichens, and patches of scrub oak and pitch pine.

The Massachusetts Natural Heritage and Endangered Species Program recognizes two distinct types of coastal heathland, the dune heathland community, which has many species in common with the barrier beach and dune community, and the glacial deposit upland, or glacial plain, community, typically represented by more inland successional heathlands. Coastal heathland is also closely associated with sandplain grassland.

Our best heathlands contain remarkable aggregations of rare plants, insects, and birds, including many showy species. The Nantucket moors still hold breeding pairs of the State Threatened northern harrier and State Endangered short-eared owl. The heathland of Naushon and Nashawena islands provides one of the last major New England strongholds of the threatened grasshopper sparrow.

The wild, slightly melancholy beauty of our most extensive coastal heathlands has much in common with the brooding splendor of British moorland, whether the Dartmoor of *The Hound of the Baskervilles* or the Yorkshire moors of *Wuthering Heights.* And in fact our most extensive remaining heathlands—on Nantucket—are known affectionately on that island as "The Moors." But whereas European moors consist mainly of various heather species, American heaths contain no heather at all and are typically dominated by their close relatives blueberry and huckleberry.

Heathlands display a remarkable spectrum of seasonal color and mood. In early spring the new pink and pale green leaves and the white blossoms of shadbush hanging above bell-flowered bearberry and lowbush blueberry, all floating in coastal mist, make for a gentle pastel awakening of the new season. After Memorial Day carpets of bird's-foot violet and splotches of golden heather (not true heather or even a heath) give way in midsummer to white flat-topped aster and wood lilies. September brings an impressive flowering of

Previous Page: Heathland, Nantucket

Grasshopper sparrow

lavender asters and complementary goldenrods. A few weeks later the huckleberry leaves go up in a blaze of scarlet. Even in winter there is a fine array of sensible earth tones—the mineral green of *Cladonia* lichens, the rusty dried stems of bushy rockrose—showing through patchy snow.

No sensory description of heathland is complete without reference to its distinctive smell, a spicy aroma emanating from the volatile oils in huckleberry, bayberry, wintergreen, and other plants.

INDICATOR SPECIES

❧ TREES: None or occasional individual pitch pines or scrub oak. ❧ SHRUBS (HEATHS): Early lowbush blueberry, late low-bush blueberry, black huckleberry, trailing arbutus, bearberry, sheep laurel, wintergreen. ❧ SHRUBS (NONHEATHS): Broom crowberry, Nantucket shadbush (SSC), bayberry, golden heather, false heather, chokeberry. ❧ GRASSES AND SEDGES: Little bluestem (grass), Pennsylvania sedge. ❧ WILDFLOWERS: Bird's-foot violet, eastern silvery aster (SE), showy aster, stiff-leaved aster, purple cudweed (SE), three-toothed cinquefoil (Cape Cod north), bushy rockrose (SSC), downy goldenrod, seaside goldenrod. ❧ GRASSHOPPERS: Huckleberry locust. ❧ BUTTERFLIES AND MOTHS: Regal fritillary (SE, now extirpated; caterpillar feeds on bird's foot and other violets), brown elfin (blueberries and huckleberries), Leonard's and dusted skippers (larvae of both feed on bluestem grasses), chain-dot geometer (SSC), coastal heathland cutworm

KEY
FE=Federally Endangered
SE=State Endangered
ST=State Threatened
SSC=State Special Concern
O=Obligate

Eastern silvery aster

(SSC), barrens buckmoth (ST; scrub oak). ☙
BEETLES: Purple tiger beetle (SSC). ☙ **REPTILES:** Hognose snake (not on the islands), smooth green snake. ☙ **BIRDS:** Raptors are frequent at all seasons owing to the abundance of meadow voles. Species include northern harrier (ST), red-tailed hawk, rough-legged hawk (winter), heath hen (extirpated), peregrine falcon (FE; migrant), merlin (migrant), American kestrel, short-eared owl (SE). Other typical heathland birds are whimbrel (fall migrant), American golden-plover (fall migrant), grasshopper sparrow (ST), vesper sparrow (ST), savannah sparrow. ☙ **MAMMALS:** Meadow vole, short-tailed shrew.

DISTRIBUTION

Heathlands occur worldwide, especially at high latitudes and altitudes and in coastal areas where harsh climatic and growing conditions favor hardy ericaceous plants and limit other vegetation types. New England coastal heathland ranges from Nova Scotia to northern New Jersey. The most diverse and extensive examples are found in Massachusetts: on Cape Cod, Martha's Vineyard, Nantucket, and the Elizabeth Islands.

EXTENT IN MASSACHUSETTS

Approximately 4,000 acres in the central moors of Nantucket. Other significant sites occur at Wasque on Martha's Vineyard, near Marconi Station, Wellfleet, and in scattered localities along the coast of Maine and in the Canadian Maritime Provinces.

CONSERVATION STATUS

European colonists remarked on the distinctive presence of coastal heathlands and rapidly proceeded to decimate one of its endemic birds, the heath hen; the last individual of this subspecies of the greater prairie chicken was seen on the heathlands and scrub oak barrens of Martha's Vineyard on March 11, 1932.

The new settlers were also responsible for increasing the acreage of New England heathland by converting the forests to cropland and pasture. Once farming and grazing ceased, areas with the poorest soils, especially along the coast, succeeded to heathland. Most of the Nantucket moors were used as sheep pounds until the mid-nineteenth century. By the late 1900s the Nantucket sheep commons were being referred to as moors— doubtless to appeal to the romantic tastes of the burgeoning numbers of tourists increasingly drawn to the island.

The Nantucket moors were also burned to attract the large flocks of American golden-plovers, whimbrels, and Eskimo curlews that stopped here in late summer and fall, especially in stormy weather, to feed on berries and insects en route from their Arctic breeding grounds to their South American wintering grounds. By midcentury a thriving business had developed around the killing of these birds for the market. Fred Bodsworth, the author of *The Last of the Curlews* writes: "Often they alighted on Nantucket in such numbers that the shot supply of the island would become exhausted and the slaughter would have to stop until more shot could be obtained from the mainland." By the turn of the century shorebird species that had once occurred by the tens of thousands were scarce or absent altogether. Thanks to the conservation laws promoted by the Massachusetts Audubon Society, such reckless slaughter became illegal, but the species involved never regained their former numbers, and the Eskimo curlew remains on the brink of extinction.

More recently, vast areas of coastal heathland have been destroyed for housing and associated development such as shopping malls. Most remaining New England coastal heathlands are managed by conservation organizations, mainly using prescribed burning to maintain them in a biologically diverse mosaic of successional stages. However, even prime areas of the Nantucket moors continue to be converted to golf courses and similar uses. (See also SANDPLAIN GRASSLAND.)

PLACES TO VISIT

Sesachacha Heathland Wildlife Sanctuary, Nantucket. Massachusetts Audubon Society.

Altar Rock, Central Moors, Nantucket. Nantucket Conservation Foundation.

Wasque, Chappaquiddick, Martha's Vineyard. The Trustees of Reservations.

Marconi Station, Cape Cod National Seashore, South Wellfleet, National Park Service.

FURTHER READING

Changing Landscapes: A Pictorial Field Guide to a Century of Change on Nantucket, by Peter W. Dunwiddie. Nantucket, Mass.: Nantucket Historical Association, 1992.

"The Heath Hen," in Alfred O. Gross, *Memoirs.* Boston Society of Natural History (1928), vol. 6, no. 4, pp. 491–588.

The Last of the Curlews, by Fred Bodsworth. New York: Dodd, Mead & Company, 1954. Reissued by Buccaneer Books, 1993.

Martha's Vineyard Landscapes: The Nature of Change, by Peter W. Dunwiddie, Vineyard Haven, Mass.: Vineyard Conservation Society, 1994.

These Fragile Outposts, by Barbara Blau Chamberlain. Yarmouth Port, Mass.: Parnassus Imprints, 1981.

Early lowbush blueberry

Monarch on rough-stemmed goldenrod

Pitch Pine–Scrub Oak Barrens

Pitch pine

Cape Cod is only a headland of high hills of sand, overgrown with shrubby pines, hurts [huckleberry], and such trash.

CAPTAIN JOHN SMITH, CA. 1602

*L*ike sandplain grasslands and coastal heathlands, this community occurs on the very dry, acidic, nutrient-poor glacial sand deposits found mainly on our southeastern coastal plain, in the Connecticut River Valley, and in isolated, sporadically distributed patches elsewhere. A similar community that contains many of the same features grows in areas with exposed ledge and thin soil, such as Cape Ann, Middlesex Fells, and the Blue Hills. The characteristic gestalt of New England pine-oak barrens is an open growth of small pines, with a dense, 7-to-10-foot-high understory of scrub oaks and 3-foot-high huckleberry. This formation may thrive in relative stability in places where living is a little easier than on the harshest coastal sands, but not easy enough to nurture a pine-oak woodland. Or such barrens may be a successional community, a robust shrubland in transition from heath to forest.

The most fascinating trait of the pine-oak barrens is its complex relationship with fire. Like the mythical Phoenix, it must burn in order to reproduce. The naturally dry habitat is prone to fire from the outset. Once a fire starts, the volatile oils present in many of the barrens plants, such as pitch pine and bayberry, feed the flames. The woody species have evolved effective means of surviving the conflagration, such as thick bark or the ability to resprout quickly from their root crowns.

But these plants don't just survive the fire, they actually thrive on it. Pitch pine produces special cones that give up their seeds only in intense heat, and the seeds of barrens flowers that have lain dormant for years in the leaf litter must be well scorched in order to germinate. In place of the decomposition of leaf litter that creates fertile soil in moister communities, a concentrate of nutrient-rich ash is created by the fire and gives the sterile sands a charge of fertilizer to encourage the new growth. The new plant growth attracts other life forms seeking the new leaves, nectar, and fruits. Ants carry crops of seeds off to their nests; birds pass fruits through their guts; and butterflies, moths, bees, and flies distribute pollen and find new nectar sources, spreading the reproductive stimulus more broadly. Research on this network of fire-stimulated activity has clearly demonstrated that the natural biological diversity explodes in the barrens in the period just following a good burn and then gradually declines as the community matures and the scrub oaks begin to overshadow other species.

Temperature also affects the growth pattern of this community. In many cases, barrens correspond with frost pockets, low areas in the landscape with a colder micro-climate that inhibits succession.

Looking at the dense pine, oak, and huckleberry that smother the rolling sandplains in places such as Myles Standish State Forest in Plymouth, few people would think to call it "barren"—even less so on a summer morning with a chorus of prairie warblers, field sparrows, and eastern towhees on the air and a host of multicolored butterflies winging about in search of nectar and egg-laying sites—or on a warm, still evening with a thunderstorm approaching, whip-poor-wills wailing, and dozens of boldly marked moths fluttering into a blacklight. In fact, this habitat at its best holds an impressive concentration of the Commonwealth's rarest species and is a great playground for curious naturalists—providing that you take an intelligent approach to your explorations. If you attempt to crash wantonly through the impenetrable, tick-infested scrub at high noon, the barrens will grind you up and send you back to your car scratched, parched, and miserable.

Eastern towhee in scrub oak

INDICATOR SPECIES

❧ **TREES:** Pitch pine, scrub oak, dwarf chestnut oak (more common on inland sites). ❧ **SHRUBS:** Black huckleberry, early lowbush blueberry, late lowbush blueberry, bearberry, trailing arbutus, broom crowberry. ❧ **GRASSES AND SEDGES:** Little bluestem (grass), Pennsylvania sedge. ❧ **WILDFLOWERS:** Cow-wheat, wild lupine, hairy wild lettuce, lion's foot, aromatic boneset. ❧ **BUTTERFLIES AND MOTHS:** Edward's hairstreak (caterpillar feeds on scrub oak), hoary elfin (bearberry), pine elfin (pitch pine), sleepy duskywing, Horace's duskywing (oaks), Persius duskywing (ST, obligate species; very rare and local; wild lupine). Karner blue, a subspecies of the Melissa blue, has never been confirmed as breeding in Massachusetts, but has been recorded near Albany and in southern New Hampshire, and could well have occurred in the barrens of the Connecticut Valley before they were destroyed; the larva feeds on wild lupine. Barrens daggermoth (ST), blueberry sallow (ST), Gerhard's underwing (ST), barrens buckmoth (ST), Bucholz's gray (ST), pine barrens itame (SSC), barrens metarranthis (SSC). ❧ **AMPHIBIANS:** Fowler's toad, gray tree frog. ❧ **REPTILES:** Eastern box turtle (SSC), hognose snake, worm snake (ST; Connecticut Valley). ❧ **BIRDS:** Heath hen (extinct), ruffed grouse, northern bobwhite, whip-poor-will, common nighthawk, fish crow, eastern bluebird, brown thrasher, hermit thrush, Nashville warbler, pine warbler, prairie warbler, eastern towhee.

KEY
ST=State Threatened
SSC=State Special Concern

Barrens buckmoth on bayberry

DISTRIBUTION

The pitch pine–scrub oak community described here ranges from southern Maine to Long Island and west to New York, and within this area, distinctions can be made between northern, inland, and coastal variants. The best examples in Massachusetts are in Plymouth County, on Cape Cod, and on Martha's Vineyard.

EXTENT IN MASSACHUSETTS

According to *Forest Statistics for Massachusetts— 1975 and 1985* (Dickson and McAfee, 1988, USDA Forest Service), there were 103,100 acres of pitch pine forest in Massachusetts in 1985, or about 3.5 percent of all forested lands. This is not equivalent to pine-oak *barrens,* for which careful estimates are difficult to find, but may have some relevance at the order of magnitude level. The greatest acreage of Massachusetts pitch pine–scrub oak barrens is concentrated in Plymouth County, followed by Cape Cod, Martha's Vineyard, and Nantucket. The community was once extensive on the sand-filled bed of the ancient glacial lake Hitchcock in the Connecticut River Valley, but inland examples such as this are now all but destroyed by development.

CONSERVATION STATUS

Pitch pine–scrub oak barrens are threatened mainly by housing development and fire suppression. In Plymouth County and on Cape Cod, vast areas were subdivided during the housing booms of 1970s and 1980s; the once extensive barrens in the Westfield, Chicopee, and Springfield areas are all but lost to development. As noted, this community requires fire to achieve its full biological potential, and human control of this natural element is probably implicated in the extirpation of a number of species, including the heath hen and the karner blue butterfly. The coastal plain has also been subjected repeatedly to extensive aerial spraying of pesticides to kill mosquitoes and gypsy moths, which may have reduced populations of the very rich lepidopteran fauna of the barrens. While extensive areas of pitch pine–scrub oak barrens have some measure of protection within state parks, they are rarely managed properly for their ecological value and in many cases are being fragmented and otherwise degraded by unrestricted recreational uses.

PLACES TO VISIT

Myles Standish State Forest, Plymouth. Massachusetts Department of Environmental Management (DEM).

Manuel F. Correllus State Forest, Edgartown. DEM.

Sesachacha Heathland Wildlife Santuary, Nantucket. Massachusetts Audubon Society.

Nickerson State Park, Brewster. DEM.

FURTHER READING

"Element Stewardship Abstract for Northeastern Pitch Pines/Scrub Oak Barrens," Arlington, Va.: The Nature Conservancy, n.d.

"The Massachusetts Pine Barrens," *Sanctuary* (Massachusetts Audubon Society), September– October 1993.

Oak-Conifer Forest

This whole forest consisted of some two or three hundred enormous oaks and ash trees. Their graceful, mighty trunks showed magnificently dark against the aureately translucent greenery of the hazel bushes…brightly colored woodpeckers tapped hard at their thick bark…a reddish brown squirrel would leap sprightly from tree to tree…[and] under the light shade of the beautiful chiseled fronds of the lady fern, violets and lilies of the valley blossomed; agarics, oak truffles, crimson toadstools— all grew there….

IVAN TURGENEV
The Hunting Sketches

People who study forests pretty much agree that between the spruce-fir and northern hardwood forests of northern New England and the oak-hickory– yellow poplar woodlands that start to dominate from central Connecticut south, a mixed forest prevails, dominated by northern red oak, white pine, eastern hemlock, red maple, and American beech and often containing many other tree species such as black and white birches, black cherry, white ash, and sugar maple. There is little agreement, however, on what this forest is *called.*

Among many names for the forest type prevailing in most of Massachusetts are hemlock–white pine–northern hardwood forest, white pine– northern red oak–red maple forest, Appalachian oak forest, and mesic/dry acidic oak-conifer forest. It is often spoken of as a "transitional" forest, and it is frequently divided into regional variations or distinct forest types, depending on which tree species is dominant. Some authorities split off an "Appalachian oak" or "oak-hickory" forest, which begins in the southeastern extremity of Maine and covers the milder eastern and southern portions of the Commonwealth. But this distinction is far easier to see on a forestry map than

walking in the woods in eastern Massachusetts. For present purposes, oak-conifer forest means the basic, common, highly variable, upland forest type of Massachusetts, covering the low hills and rolling peneplain from the pine barrens of the sandy coastal plain to the edge of the rich, cold western highlands.

Typically, oak-conifer forest grows on well-drained, nutrient-poor, relatively thin soils (8–15 inches deep) over acidic bedrock such as granite or gneiss. The oak, pine, and hemlock leaves decompose slowly, creating deep leaf litter and contributing their own acidity to the soil chemistry.

Another pervasive influence, as important as soil and climate in creating the oak-conifer forest that covers most of Massachusetts today is "us." By 1840, the European colonists had felled these forests almost completely, often repeatedly, for agriculture, firewood, and lumber profits. Many of the settlers then departed—to the more fertile west, to urban factories—allowing the forest to recreate itself as a reflection of the preceding land use patterns. In the precolonial forest, white pines probably appeared mainly as lone giants, destined to hold the sails of the King's navy. Today a grove of white pine often stands where an abandoned

Opposite: Oak-conifer forest, Pleasant Valley Wildlife Sanctuary, Lenox

farmstead or other clearing allowed this sun-loving tree to germinate densely and "poison" the ground against other trees with its sour needles. These pine groves eventually give way as they decline to more mixed stands dominated by hardwoods.

The stature of the oak-conifer forest at maturity averages 80 to 100 feet but the tallest white pines in this forest today exceed 150 feet. The canopy is generally fairly open, allowing a varied understory of striped maple, witch hazel, sheep and mountain laurels, blueberries and other shrubs, and a ground-cover mosaic of woody species, wildflowers, ferns, and clubmosses. As with the tree species, the dominant understory and ground-cover plants vary according to soil moisture and chemistry and shade tolerance.

Because of this habitat's transitional nature, few if any species are confined to it and it does not contain concentrations of rarities, but the oak-conifer forest more than makes up for its fuzzy definition with the great breadth of plant and animal species it contains and the splendid variations it plays on its own theme. It is an inclusive rather than exclusive community, and will reward the visitor with some glimpses of wild action or aesthetic pleasure in every season: open, bright, and strewn with flowers in spring; fragrant, loud with birdsong in summer; colorful in autumn when the northwest wind whistles through the pines and hemlocks; and as still and solemn as a cathedral nave in winter.

White-tailed deer

INDICATOR SPECIES

These are typical forest species of Massachusetts. Most are common and widespread and are not confined to the oak-conifer community but occur in other types of forests as well. More restricted forest species are noted under the other forest communities described. ☙ CANOPY TREES: Red oak (D), white pine (D), eastern hemlock (D), red maple (D), American beech (D), black cherry, black birch, white birch, black oak, white ash, American basswood, sugar maple, shagbark hickory, pignut hickory, bitternut hickory, sweet pignut hickory, mockernut hickory. ☙ UNDERSTORY TREES, SHRUBS, AND VINES: American chestnut (formerly a common canopy species, now blighted by an Asiatic fungus), witch hazel, striped maple, mountain laurel, sheep laurel, common or thicket shadbush, lowbush blueberry, flowering and roundleaf dogwoods, American fly honeysuckle, pink and mountain azaleas, maple-leaf viburnum, hobblebush, highbush cranberry, prickly and other gooseberries, Virginia creeper. ☙ FERNS AND CLUBMOSSES: Marginal wood fern, spinulose wood fern, oak fern, broad beech fern, long beech fern, New York fern, ebony spleenwort, silvery spleenwort, lady fern, hay-scented fern, Christmas fern, common polypody, interrupted fern, cinnamon fern, daisy-leaved grape fern, cut-leaved grape fern, rattlesnake fern, ground cedar, running pine, shining clubmoss, staghorn clubmoss, stiff clubmoss, tree clubmoss. ☙ GRASSES, SEDGES, AND RUSHES: Pennsylvania sedge, laxiflora sedge, bottlebrush grass, long-awned wood grass, wood reedgrass, wood rush. ☙ WILDFLOWERS: Most common and characteristic species are trout lily trillium, nodding trillium, large-flowered trillium, purple trillium, painted trillium, bloodroot, common and violet wood-sorrels, jack-in-the-pulpit, common and sessile-leaved bellworts, great Solomon's-seal, hairy Solomon's-seal, false Solomon's-seal, twisted stalk, yellow clintonia, wood and rue anemones, May-apple, foam flower, one-sided pyrola, round-leaved pyrola, green-flowered pyrola, shinleaf, goldthread, hepaticas, swamp dewberry (vine), starflower, white and red baneberries, smooth and downy false foxgloves, pipsissewa, spotted wintergreen, sweet white violet, large-leaved violet, northern blue violet, downy yellow violet, smooth yellow violet, beechdrops (a parasite that grows on beech roots), Indian pipes (saprophyte), pinesap (saprophyte), wild columbine, Canada mayflower, pink lady's-slipper, downy rattlesnake-plantain, green wood orchis, helleborine, whorled-pogonia, small whorled-pogonia (FE; one of only three Massachusetts plant species on the federal endangered species list; restricted to oak-conifer forest), hairy and downy wood mints, bunchberry tick-trefoils, wood lousewort, trailing arbutus, cow-wheat, wild sarsaparilla, partridgeberry, wintergreen Indian cucumber root, sharp-leaved goldenrod, blue-stemmed goldenrod, large-leaved goldenrod, slender goldenrod, large-leaved aster, heart-leaved aster, whorled aster, white wood aster. ☙ DRAGONFLIES: Ringed bog haunter (adults in woods near cold acid bogs or ponds). ☙ BUTTERFLIES: Tiger swallowtail (caterpillar feeds on many tree species), spicebush swallowtail (sassafras), eastern pine elfin (pines), banded hairstreak (oak), hickory hairstreak (hickory), comma (elm, hops), gray comma (currants), Compton's tortoiseshell (birch), mourning cloak (willow),

KEY
D=Dominant
FE=Federally Endangered
SE=State Endangered
SSC=State Special Concern

red-spotted purple (cherry), white admiral (birch), northern pearly eye (grasses), little wood satyr (sedges), dreamy duskywing (willow, poplar), Juvenal's duskywing (oaks). ◡ **AMPHIBIANS:** Red eft (terrestrial form of red spotted newt), northern dusky salamander (stream edges and other moist areas), redback salamander, American toad, gray tree frog, wood frog; 6 species of mole salamanders occur in woodland pools (see Amphibians, VERNAL POOL). ◡ **REPTILES:** Eastern box turtle (SSC), northern redbelly snake, northern ring-neck snake, northern black racer, black rat snake (SE), eastern milk snake, northern copperhead (SE), timber rattlesnake (SE). ◡ **BIRDS:** Turkey vulture, Cooper's hawk, northern goshawk, broad-winged hawk, red-tailed hawk, ruffed grouse, wild turkey, mourning dove, passenger pigeon (extinct owing to commercial and recreational hunting), black-billed cuckoo, yellow-billed cuckoo, eastern screech-owl, great horned owl, whip-poor-will, red-bellied woodpecker, yellow-bellied sapsucker, downy woodpecker, hairy woodpecker, northern flicker, pileated woodpecker, eastern wood pewee, great crested flycatcher, blue jay, American crow, fish crow, black-capped chickadee, tufted titmouse, red-breasted nuthatch (conifers), white-breasted nuthatch, brown creeper, winter wren, hermit and wood thrushes, American robin, solitary vireo, yellow-throated vireo, red-eyed vireo, chestnut-sided warbler, black-throated blue warbler, black-throated green warbler (conifers), Blackburnian warbler (conifers), pine warbler (white pine), black-and-white warbler, American redstart, worm-eating warbler, ovenbird, scarlet tanager, rose-breasted grosbeak, dark-eyed junco, brown-headed cowbird (brood parasite), Baltimore oriole, purple finch. ◡ **MAMMALS:** Opossum, masked shrew, short-tailed shrew, silver-haired bat, eastern pipistrelle, red bat, hoary bat (other bat species may feed at forest edges but are not essentially forest-dwelling animals), New England cottontail, snowshoe hare, eastern chipmunk, gray squirrel, red squirrel (prefers conifers), southern flying squirrel, northern flying squirrel, white-footed mouse, deer mouse, red-backed and pine voles, woodland jumping mouse, porcupine, gray fox, black bear, raccoon, fisher, long-tailed weasel, mink (near water), bobcat, moose, white-tailed deer, eastern timber wolf (formerly abundant, now extirpated), eastern mountain lion (extirpated).

DISTRIBUTION

Although oak-conifer forest is the dominant forest community of Massachusetts, as described here, it has relatively restricted distribution in North America. It occurs sporadically from the Great Lakes states east to southern Ontario, western New York, and central Pennsylvania and in patches farther south in the Appalachians. In New England it ranges from southwestern Maine (near latitude 45 degrees north), across southern New Hampshire and along the Connecticut River in eastern Vermont south to central Connecticut and southernmost Rhode Island. With the exception of the pitch pine–scrub oak barrens of Cape Cod and the northern hardwood forest and spruce-fir forest of the western highlands, this forest prevails throughout the uplands of the Commonwealth.

EXTENT IN MASSACHUSETTS

In Thoreau's day, about 75 percent of Massachusetts had been cleared of forest. Shifting economic priorities and the consequent changes in human behavior have allowed the native oak-conifer forest to return, though in an altered state, and the situation has now reversed, with about 64 percent of the state covered in various forms of second-growth forest. As defined here (that is, including oak-hickory, oak-pine, and hemlock associations), Massachusetts oak-conifer forest covers about 21,942,000 acres or about 66 percent of our total forest cover (Dickson and McAfee, 1988).

Opposite: Timber rattlesnake

Worm-eating warbler

CONSERVATION STATUS

Of the 3,225,200 acres of forest in Massachusetts, 2,929,400 acres (91 percent) are "timberland," defined by the USDA Forest Service as "forest land producing or capable of producing crops of industrial wood and not withdrawn from timber utilization.…Formerly known as commercial forest land." Only 101,400 acres (3 percent) are "productive reserved," that is, forest land sufficiently productive to qualify as timberland, but withdrawn from timber utilization through "statute, administrative designation, or exclusive use for Christmas tree production." (All quotes and statistics are from Dickson and McAfee, 1988.) Virtually all of these forests are second growth, many of them relatively young and/or even-aged because they have been logged in the recent past. The majority of our forests, even those on conservation land, are "managed," meaning that they are logged on a regular basis. In addition to the obvious economic value, timbering is said by traditional foresters to enhance "wildlife values." This generally means the creation of openings and scrub (nonforest habitat) favored by game species such as white-tailed deer and ruffed grouse. As the preceding figures show, mature, undisturbed forest is a great rarity in Massachusetts and, indeed, anywhere in North America. Therefore, Massachusetts Audubon has articulated a forest-management policy on its sanctuaries calling for the maintenance and, where possible, creation of large, unfragmented, undisturbed tracts of forest that will be allowed to mature and regenerate indefinitely. No removal of trees is permitted on the forest reserves within our sanctuary system.

Besides timber management, other threats to oak-conifer forest and its inhabitants are

- Exotic pest insects, especially gypsy moth, hemlock looper, and hemlock wooly adelgid.
- Invasion by aggressive, invasive exotic plant species such as European buckthorn, Oriental bittersweet, and Japanese barberry.
- Fragmentation by roads and development, resulting in the creation of forest "islands" that are too small to support reproductively viable populations of many species.
- Excessive brood parasitism of songbirds and predation of nestling birds and other small forest organisms by brown-headed cowbirds, crows, jays, grackles, raccoons, and other species that thrive in edge habitats and have increasing access to forests as a consequence of fragmentation.
- Ozone damage to white pines.
- Increasing abuse and disturbance by trail bikes and other off-road vehicles.

PLACES TO VISIT

The majority of conservation and recreation lands in Massachusetts contain examples of oak-conifer forest. A sampling of the most extensive and accessible examples is as follows:

Moose Hill Wildlife Sanctuary, Sharon. Massachusetts Audubon Society (MAS).
Broadmoor Wildlife Sanctuary, South Natick. MAS.

Wachusett Meadow Wildlife Sanctuary, Princeton. MAS.

Wachusett Mountain State Reservation, Princeton. Massachusetts Department of Environmental Management.

Brooks Woodland Preserve, Petersham, The Trustees of Reservations (TTOR).

Monument Mountain Reservation, Great Barrington. TTOR.

FURTHER READING

Changes in the Land, by William Cronon. New York: Hill & Wang, 1983.

Deciduous Forests of Eastern North America, by E. Lucy Braun. New York: Macmillan, 1950.

Eastern Forests, by John Kricher and Gordon Morrison. Boston: Houghton Mifflin, 1988.

A Field Guide to Trees and Shrubs, by George A. Petrides. Boston: Houghton Mifflin, 1958.

Forest Cover Types of the United States, edited by F. H. Eyre. Washington, D.C.: Society of American Foresters, 1980.

Forest Statistics for Massachusetts—1972 and 1985, by David R. Dickson and Carol L. McAfee. Broomal, Pa.: Northeastern Forest Experiment Station (USDA Forest Service), 1988.

A Natural History of Trees of Eastern and Central North America, by Donald Culross Peattie. Boston: Houghton Mifflin, 1948; reissued 1991.

The North Woods: An Inside Look at the Nature of Forests in the Northeast, by Peter T. Marchand. Boston: Appalachian Mountain Club, 1987.

Stone Walls and Sugar Maples: An Ecology for Northeasterners, by John Bank and Marjorie Holland. Boston: Appalachian Mountain Club, 1979.

Trees of North America: A Guide to Field Identification, by C. Frank Brockman. New York: Golden Press, 1988.

Redback salamander

Vernal Pool

Vernal pool in early spring

Blunt-lobed hepatica

The frog does not drink up the pond in which it lives.

NATIVE AMERICAN PROVERB

For the purpose of legal protection, vernal pools are defined as confined basin depressions that in most years contain water for at least two continuous months in spring and/or summer, that lack fish, and that are located in areas protected under federal, state, or local wetlands law. But definitions give little hint of the magical qualities of these transitional habitats, which serve as a natural gateway between the terrestrial and aquatic realms. Nor do they begin to do justice to the biological richness of these pools, which during the height of spring activity inspire equally rich descriptors such as swarming, teeming, throbbing, and pullulating.

Vernal pools, also known as temporary or ephemeral ponds or pools, occur in a variety of situations. They may be associated with other aquatic habitats such as rivers and lakes or they may fill the bottom of glacial kettleholes in the midst of upland forests. Vernal pools may fill with late autumn rains and remain frozen through the winter, or they may stay dry until spring when snow melt and runoff raise the water table.

The critical characteristic of vernal pools is isolation, in both space and time. Because they are as a rule available as aquatic habitat only for a short period of the year and are not directly connected to large permanent bodies of water, they can be inhabited only by organisms specially adapted to limiting factors such as long dry periods and low oxygen levels. These conditions rule out fish and other large predators, whose absence renders vernal pools relatively safe and thus suitable for unrestrained reproduction by certain characteristic species. The species that have evolved in association with this isolated habitat tend to become isolated themselves, so that vernal pools contain many obligate species, organisms that are unable to complete their life cycles without this habitat. This makes vernal pools particularly valuable— and vulnerable—as wildlife habitat.

The energy flow and food web in a vernal pool are very different from those in a permanent pond. In a pond, life is based on the conversion of the sun's energy by algae and other green plants into food for the animal communities. All of the

Marbled salamander

animals return nutrients as wastes, which in turn promotes more plant growth. The vernal pool system is based on detritus. Dead leaves from the surrounding forest fill the pool bottom, where they are decomposed by fungi and bacteria and broken down into small pieces by living shredders such as caddisflies and isopod crustaceans. The smallest fragments are suspended in the water to be eaten by zooplankton (tiny animals that live adrift in the water column), fairy shrimp, and tadpoles; larger pieces are consumed on the bottom by insects, mollusks, worms, and crustaceans. The detritus feeders are eaten in turn by other organisms such as water beetles and salamander larvae.

The annual *rhythm* of life in a vernal pool is also quite different from that of a pond. The most dynamic period in the cycle is early spring, when the rising water level triggers the emergence of the many species that have lain dormant through the "dry season." As early as the first weeks of March, spring rains during the night signal the emergence from the ground of several species of mole salamanders, which follow genetically programmed routes overland to the pools to breed. The peak of this vernal orgy is announced by the distinctive quacking of wood frogs, another obligate species, beginning around mid-March. For a short period in March or April an active pool may be quite crowded with adult amphibians and their egg masses as well as colorful fairy shrimp and a host of less conspicuous bottom-dwelling invertebrates. Lest the word "crowded" be underappreciated, a single one-acre pool in Concord, Massachusetts, has been found to contain, in a single night, as many as 3,500 wood frogs, 1,500

spring peepers, 200 American toads, and 500 spotted salamanders, to name only the most numerous species; over a period of a month, the total number of these species reached over 15,000 (Brian Windmuller, personal communication).

As the pool starts to dry out with the approach of summer, many pool organisms must complete their life cycles before low oxygen and lack of moisture make the pools uninhabitable. Frogs and salamanders complete their egg and larval stages rapidly and crawl back into the forest and underground; aquatic insects such as caddisflies and dragonflies complete their pupal or nymphal stages and emerge as flying adults. Algal and duckweed mats often form on the surface and the diminishing puddle may be crammed with water beetles and bugs. Worms, mollusks, and eggs of many invertebrates remain in the mud and can withstand even severe dryness in a dormant state. But many pool beds remain damp and nutrient-rich and may be partially overgrown with herbaceous plants typical of swamps and river bottoms. During fall rains adult marbled salamanders arrive at the dry pond depressions to breed and the invertebrate community begins to come back to life. Once the pool is reflooded, the larval marbled salamanders and other organisms remain active, in many cases under the ice throughout the winter.

During the day, vernal pools are easily dismissed as mere forest rain puddles, and few people of sound mind wander of their own free will out into the woods on the rainiest nights of March. Consequently, these thriving, diverse communities are one of nature's best-kept secrets, rather resembling the villages of elves that children in fairy tales discover accidentally. As naturalists have learned more about them, however, vernal pools have developed something of a cult following. Now knots of fanatics in slickers, Wellingtons, and even wetsuits can be seen throughout the Commonwealth on the nastiest nights of the year escorting migrating salamanders through tunnels specially built to give them safe passage under busy roads, watching the sperm-drop ritual that passes for sex among salamanders, and recording statistics on the pools so that they may be legally protected. Many who have been cajoled into spending a spring night engaged in this bizarre ritual claim that the experience changed their lives—for the better!

INDICATOR SPECIES

Included are obligate species, which depend for their existence on vernal pools, and facultative species, which strongly favor vernal pools but also inhabit permanent ponds. Not listed are the wide range of wetland plants and animals that may visit or even breed in vernal pools from time to time but are not characteristic of the community. There are no true indicator plants in this community. ❧ **TREES AND SHRUBS:** Pools are often fringed with species typical of wooded swamps such as red maple, buttonbush, highbush blueberry, swamp azalea, and sweet pepperbush. ❧ **FERNS:** Royal, sensitive, cinnamon, and other water-tolerant fern species are facultative. ❧ **GRASSES, SEDGES, AND RUSHES:** Typical facultatives include brown beaksedge, sooty beaksedge, threeway sedge, mud-fruited rush, several species of manna grasses and panic grasses. ❧ **WILDFLOWERS:** A wide variety of wetland species may occur along the edges or sprout from the pool bottom as it dries out in summer. Aquatic plants such as duckweeds, skunk cabbage, bladderworts, St. John's-worts and water-crowfoots, which prefer shallow, stagnant water, wet meadows, or marshes, are also frequent. ❧ **OLIGOCHAETE WORMS:** A number of species, including some obligates. ❧ **FLAT WORMS:** Several common species, including obligates. ❧ **MOLLUSKS:** Most are facultative. Swamp fingernail clam, pond fingernail clam,

---

KEY
O=Obligate
SE=State Endangered
ST=State Threatened
SSC=State Special Concern

ubiquitous pea clam (semi-obligate), Herrington's fingernail clam, polished tadpole snail, flat-coiled gyraulus, keeled promenetus, *Physa* species. ❧ **MAYFLIES:** Typical primitive minnow mayfly, species of flatheaded mayflies, early brown spinner, intermediate spinner (floodplain pools), Borcher's drake. ❧ **DRAGONFLIES:** All are facultative. Emerald dryad and possibly other spreadwing damselflies, spatterdock or spring blue darner (SE; possibly obligate; the only spring flying *Aeshna* in New England), petite emerald, yellow-legged (and perhaps other) meadowfly species. ❧ **CADDISFLIES:** All are facultative. Cigar-tube case makers, log-cabin case makers. ❧ **TRUE BUGS:** Pond species of giant water bugs, backswimmers, water boatmen, and water striders are found in some vernal pools. ❧ **BEETLES:** Crawling water beetles, diving beetles. Whirligig beetles and water scavenger beetles are pond species that fly into vernal pools after they have filled or melted to feed on the resident life forms. ❧ **CRUSTACEANS:** Vernal fairy shrimp (O), intricate fairy shrimp (O, SSC), Massachusetts clam shrimp (O, SSC; other species for which no common names have been coined are listed by scientific name in the index under clam shrimp). ❧ **AMPHIBIANS:** Wood frog (O), spring peeper, gray tree frog, green frog, eastern spadefoot (O, ST; breeding pools do not remain the two months necessary for legal vernal pool classification), American and Fowler's toads, four-toed salamander (SSC), red-spotted newt, spotted salamander (O), blue-spotted salamander (O, SSC), Jefferson salamander (O, SSC), marbled salamander (O, ST). ❧ **REPTILES:** Spotted turtle (SSC), painted turtle, snapping turtle, wood turtle (SSC). ❧ **BIRDS:** Wood duck, solitary sandpiper (migrant). ❧ **MAMMALS:** Raccoons and other wetland-associated species may visit to feed and drink.

DISTRIBUTION

Ecosystems that have evolved in conjunction with temporary ponds occur worldwide, especially in areas with marked wet and dry seasons. The most extreme examples are desert rain pools, which may remain dry for years and come to exuberant life following torrential rains. The organisms that inhabit these systems typically have very high productivity and mortality, very rapid reproductive periods, and egg or larval stages adapted to prolonged drying. Less extreme, more predictable vernal pool systems occur mainly in the temperate zone, including in higher-altitude zones in the tropics. In North America, mole salamanders, some of which depend on temporary ponds for breeding, occur mainly east of the Great Plains, north of Mexico, and south of the boreal zone, but there are a few exceptions. Four of the ten species of North American mole salamanders occur in Massachusetts.

EXTENT IN MASSACHUSETTS

Vernal pools are common throughout Massachusetts from Cape Cod to the Berkshires, though neither wood frogs nor mole salamanders, which make up a large proportion of our vernal pool fauna, occur on Nantucket or Martha's Vineyard. Estimates of the total number of pools in the Commonwealth range from 10,000 to more than 100,000, but the actual total is probably near the high end. The town of Hubbardston has at least 228 vernal pools. Many towns have fewer, some probably have more. If we take 200 as an estimated average and multiply by the number of towns in the state (351), 70,200 is the estimated total.

CONSERVATION STATUS

Despite the fact that they are critical breeding sites for many species, vernal pools are only partially protected by Massachusetts law. The Massachusetts Endangered Species Act protects only vernal pool species listed as threatened (marbled sala-

mander and eastern spadefoot) or endangered (none) and protects these only from "taking"—for example, killing. This law also has the power to protect designated *habitats,* but to date none has been designated. There is also a measure of protection for vernal pools under Title 5 (state regulations governing siting of septic systems), forest-cutting regulations, and regulations under federal wetlands-permitting programs.

The strongest instrument of vernal pool protection is the Massachusetts Wetlands Protection Act, which protects pools and land within 100 feet that fall within legally defined "wetland resource areas" (wetlands, coastal dunes, barrier beaches, flood plains, or isolated areas that flood to a certain minimum depth and volume) *and* that are certified by the Massachusetts Natural Heritage and Endangered Species Program or that are otherwise shown to support vernal pool species. To qualify for certification, the pool must

- Be a confined basin depression
- Hold water for a minimum of two continuous months during spring and/or summer for most years
- Be free of adult fish populations
- Provide breeding habitat for vernal pool amphibians or contain fairy shrimp.

Certification also requires that the location of the pool be drawn onto a U.S. Geological Survey topographical map and a tax assessors' map and that its precise location be shown by compass bearings and distances from two landmarks, as established by professional survey or an aerial photograph.

Spotted salamander

Unfortunately, only a small percentage of the state's vernal pools have been certified. Unless a vernal pool is on conservation land, or is located in a wetland resource area protected by the state's wetlands act (as many as half are not) *and* has been identified by the conservation commission or someone else who can testify at a public hearing, it is only marginally protected.

Towns can enact bylaws to protect their own vernal pools, and about a dozen have done so.

PLACES TO VISIT

Many state and municipal parks, wildlife-management areas, and private conservation lands in the Commonwealth contain vernal pools. Although finding a vernal pool on your own is the most exciting introduction to this community, most people first encounter the habitat in the company of a knowledgeable guide. The Massachusetts Audubon Society and other organizations offer many programs through their sanctuaries statewide in March and April that visit prime examples and in some cases collect certification data. Some specific protected areas with excellent pools are as follows:

Bradley Palmer State Park, Ipswich. Massachusetts Department of Environmental Management.

Cape Cod National Seashore. (Many vernal pools in Eastham, Wellfleet, and Truro.) National Park Service.

Massachusetts Audubon Society wildlife sanctuaries with vernal pools include *Ipswich River,* Topsfield; *Broadmoor,* South Natick; *Moose Hill,* Sharon; *Broad Meadow Brook,* Worcester; and *Arcadia,* Easthampton.

FURTHER READING

Certified: A Citizen's Step-by-Step Guide to Protecting Vernal Pools, edited by Elizabeth Colburn. Lincoln, Mass.: Massachusetts Audubon Society, 1989 (3rd ed.). See also *Vernal Pool Lessons and Activities* (A curriculum companion to *Certified*), by Nancy Childs and Betsy Colburn. Lincoln, Mass.: Massachusetts Audubon Society, n.d.

Pond Watcher's Guide to Ponds and Vernal Pools of Eastern North America. Lincoln, Mass.: Massachusetts Audubon Society, 1995.

Spring Pools, by Ann Davner. Boston: New England Aquarium, 1988.

Red Maple Swamp

The scarlet of the maples can shake me like a cry of bugles going by.

WILLIAM BLISS CARMAN
"A Vagabond Song"

The red maple swamp is one of the commonest and most important natural community types in Massachusetts. In the broadest sense it may be defined as a forest dominated by red maple growing from undrained, hydric (wetland) soils. These forests normally are flooded or at least thoroughly saturated for part of the year, usually winter and spring, but may be dry enough to walk through in late summer and fall. Beyond this simple description, variation is great. Wooded swamps may occur in a variety of physical settings including vast, glacial lake basins; small depressions subject to the flow of surface water; kettleholes below the water table; pond, lake, and stream margins; abandoned river oxbows or stream channels; seeping hillsides; or other situations that maintain the necessary degree of wetness. Red maples are often rooted in mucky organic soil, but they fare just as well in the sand of a wet depression in the dunes.

Structurally, a red maple swamp may contain five distinct layers of vegetation: the mature trees that form the canopy; the younger saplings; a lower shrub layer; an herbaceous stratum of ferns, wildflowers, grasses, and sedges; and a ground cover of mosses and clubmosses. The structural complexity is matched by high biological diversity, with at least 50 species of trees, 90 species or shrubs and vines, over 300 species of nonwoody plants, more than 200 species of vertebrate animals and a large (but, as usual, uncounted) number of invertebrate species. While few species are truly restricted to red maple swamps—even red maple commonly grows elsewhere—many rare and endangered species depend on these wetlands as crucial portions of their habitat.

As may be inferred from the variability just described, the red maple swamp shows many faces and moods. A young swamp or one whose growth is retarded by constant flooding will be comparatively open with small trees sprouting gracefully among sedge tussocks, creating an open, light-filled structure. A mature, fully developed forest may be more like a dark cloister with thick columns rising to 60 feet or more into the canopy (average canopy height is about 50 feet) and only dappled light reaching the floor. It is a colorful forest that undergoes striking seasonal changes. One of the first hints of spring in southern New England is a rosy suffusion of red maple buds across the lowlands that intensifies as the pink flowers and red seed clusters emerge, all before the first foliage appears in May. In summer the open swamps vibrate with birdsong and the wings of dragonflies cruising among tussocks and saplings, and many butterfly species are attracted to swamp flowers such as swamp azalea, buttonbush, swamp milkweed, and sweet pepperbush. The dark summer interior of an old swamp is lit here and there with clumps of cardinal flowers. The unique brilliance

Red maple samaras

of a New England autumn is due in no small measure to the intense crimson of red maple leaves stressed by dryness, cold, and light deprivation. By November the smooth trunks and bare branches make swathes of lavender along the horizon line. Even in the so-called dead of winter, the big scarlet fruits of winterberry holly glow in the snow-filled forest.

INDICATOR SPECIES

Because this community is so rich in plant species, an approximate total of those recorded is noted at the beginning of each group, followed by a list of the most common species that tend to *prefer* red maple swamps in southern New England and characteristic rare species with their Massachusetts legal status, where applicable. ❧ TREES: 50 species recorded. Red maple (D), white pine, eastern hemlock, Atlantic white cedar (mainly coastal), yellow birch, swamp white oak, American elm (much reduced owing to Dutch elm disease), tupelo, black ash, green ash. Rare species: Sweet bay magnolia (SE), mossycup or bur oak (SSC), slippery elm. ❧ SHRUBS AND VINES: Over 90 species recorded. Ironwood, speckled alder, spicebush, poison sumac, common winterberry, mountain holly, sweet pepperbush, highbush blueberry, swamp honeysuckle, northern arrowwood. Rare species: Swamp or fen birch (ST), great laurel (ST), bristly black currant (SSC), swamp red currant (SSC). ❧ FERNS, CLUBMOSSES, AND HORSETAILS: 26 species recorded. Spinulose

---

KEY
D=Dominant
SE=State Endangered
ST=State Threatened
SSC=State Special Concern

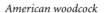

American woodcock

112

Canada warbler

wood fern, marsh fern, Massachusetts fern, netted chain fern, cinnamon fern, royal fern, ostrich fern, sensitive fern, tree clubmoss, swamp horsetail. Rare species: Hartford (or climbing) fern (SSC). ❧ GRASSES AND SEDGES: About 56 species recorded. Tussock sedge, fowl meadow grass. Rare species: Gray's sedge (ST), rigid sedge. ❧ WILD-FLOWERS: Over 200 species recorded: Skunk cabbage, false hellebore, jack-in-the-pulpit, spotted jewelweed, marsh marigold, swamp dewberry, tall meadow-rue, marsh blue violet. Rare species: Crooked-stemmed aster (SSC), meadow bitter-cress, narrow-leaved spring beauty (ST), hemlock parsley (SSC), yellow lady's-slipper (var. *parviflorum*; SE), showy lady's-slipper (SSC), broad water-leaf (SE), great blue lobelia (ST), gypsywort (ST), white adder's-mouth (ST), swamp lousewort (SE), sweet coltsfoot (var. *palmatus*; ST), lizard's tail, swamp oats (ST), Britton's violet (ST). ❧ DRAG-ONFLIES: Most woodland species require some water flow to breed. Emerald dryad, gray petaltail (very rare, possibly extirpated), spatterdock darner, shadow darner, early darner, swamp darner, harlequin darner; all species of spiketails, especially where streams are present; brown river

skimmer, ringed bog haunter, dusky emerald, Williamson's emerald. ❧ BUTTERFLIES: The following species habitually fly in open woods and use swamp plants as their larval food; the latter are listed in parentheses after each species. Henry's elfin (highbush blueberry, winterberry, European buckthorn), comma (elm), mourning cloak (willow), Appalachian brown (sedges). ❧ AMPHIB-IANS: 24 species recorded. Marbled salamander (ST), Jefferson salamander (SSC), spotted salamander, four-toed salamander (SSC), redback salamander, northern two-lined salamander, spring salamander (SSC), American toad, spring peeper, gray tree frog, wood frog. ❧ REPTILES: 18 species recorded. Eastern ribbon snake, ringneck snake, spotted turtle (SSC). ❧ BIRDS: 119 species recorded. Many widespread deciduous-forest songbirds such as blue jays, black-capped chickadees, and red-eyed vireos, nest in red maple swamps. Only species showing some *preference* for this habitat are listed. Green heron, wood duck, hooded merganser, red-shouldered hawk, barred owl, great crested flycatcher, brown creeper, veery, northern waterthrush, Canada warbler. Rare species: Acadian flycatcher (very rare breeder in Massachusetts). ❧ MAMMALS: 49 species recorded. Masked shrew, water shrew (SSC), star-nosed mole, mink, beaver.

DISTRIBUTION

Red maple is restricted to North America east of the Great Plains, from extreme southeastern Manitoba east to Newfoundland and south to eastern Texas and Florida. Red maple swamp is much more restricted, essentially being confined to the glaciated Northeast, from Manitoba through the Great Lakes states, New England, and Newfoundland south to New Jersey. Regionally, this community is most prevalent from central to southern New England. It is widespread throughout Massachusetts.

EXTENT IN MASSACHUSETTS

Massachusetts is notably rich in red maple swamp, which accounts for 6 percent of the total land area of Massachusetts and 68 percent of all our vegetated freshwater wetlands, for a total of about 300,000 acres (derived from *Ecology of Red Maple Swamps in the Northeast,* 1993).

CONSERVATION STATUS

The extensive red maple swamps of Massachusetts play a proportionately important role in water-related issues. They absorb excess precipitation and surface runoff, providing flood abatement. They store groundwater and recharge aquifers for our drinking water and other water-related activities. They clean our water supply by removing, retaining, or altering many pollutants of human origin.

Since European settlement many thousands of acres of forested wetlands have been cleared, filled, drained, or polluted in the process of providing for an increasing population and building a modern society. Although records are poor, it has been credibly estimated that Massachusetts may have lost as much as 50 percent of the wetlands that existed here at the beginning of colonization. In this century, many extensive developments such as shopping malls were sited in filled wetlands, which were regarded in less enlightened eras as wastelands. For example, in 15 communities sampled throughout the state, an average of 4 percent of the then existing vegetated wetlands were lost between 1951 to 1977. The principal causes of these losses were agricultural conversion, damming of streams for water impoundments, highway construction, and residential and commercial development. Fortunately, Massachusetts had the foresight to pass strong wetlands protection legislation, beginning in the 1960s, which provide even stricter protection than federal wetlands laws. Unfortunately, loss of these resources continues despite the laws. Many subtler phenomena of human origin are also at work. The introduction of Dutch elm disease has significantly altered the tree composition of our red maple swamps, and another aggressive alien, European buckthorn, invades the forest understory and supplants the typically diverse native shrub layer.

Despite the extensive loss of this community, Massachusetts conservationists have traditionally thought of red maple swamp as an abundant habitat and therefore until recently have given little thought to studying it thoroughly or worrying about protecting it. From a global perspective, it becomes clear that Massachusetts is a center of abundance for this spectacular and ecologically important community, and we should perhaps put additional effort into identifying the largest and most species-rich examples for strict protection. We also need to close legal loopholes that allow conversion to continue.

PLACES TO VISIT

Many protected open spaces in Massachusetts include good examples of red maple swamp. One of the largest is the *Hockamock Swamp,* which extends into several towns in southeastern Massachusetts. Access is via Pleasant Street in West Bridgewater. Massachusetts Audubon Society wildlife sanctuaries with accessible swamps include *Ipswich River,* Topsfield; *Broadmoor,* South Natick; *Moose Hill,* Sharon; *Wachusett Meadow,* Princeton; and *Pleasant Valley,* Lenox.

Other exemplary and accessible sites are:

Boxford State Forest, Boxford. Massachusetts Department of Environmental Management (DEM).

Bear Swamp, Beartown State Forest, Monterey. DEM.

Great Swamp, Whitney and Thayer Woods, Cohasset and Hingham. The Trustees of Reservations.

FURTHER READING

Ecology of Red Maple Swamps in the Glaciated Northeast: A Community Profile, by F. C. Golet et al. U.S. Fish and Wildlife Service, Washington, D.C.: 1993.

Hockamock—Wonder Wetland, Fund for Preservation of Natural Areas (n.d.; privately printed).

Freshwater Marsh

Red-winged blackbird

The Marsh Wife is sister to the Elf King and aunt to the Elf maidens…but all that people know about her is that when the mist rises over the meadows in summer, she is at her brewing. A scavenger's cart is sweet compared to the Marsh Wife's brewery. The smell from the barrels is enough to turn people faint, and wherever there is a chink between them, it is filled up with noisome toads and slimy snakes.

HANS CHRISTIAN ANDERSEN
"The Girl Who Trod on a Loaf"

Marshes are open wetlands with water shallow enough to support a dense emergent plant cover rooted in soil but deep enough to inhibit the growth of woody plants and more terrestrial herbaceous species typical of swamps. They may be flooded to a depth of six feet in the wettest seasons, but may show exposed soil during the driest period of late summer. Marsh soils are generally saturated year-round and consist of some combination of mineral and organic elements, typically with a layer of muck at the surface made up of well-decomposed plant material resting on more consolidated mineral soil below. Marshes can be relatively acid or alkaline, depending on local conditions, but are never as acid as bogs. The soil composition and chemistry reflect the fact that, unlike bogs, marshes are open systems subject to some degree of flow, which delivers minerals and flushes organics and excessive acidity.

Technical wetland classification systems distinguish among a number of freshwater marsh types with names such as "shallow emergent marsh" and "palustrine persistent emergent wetland." This account describes the broad category and the most common variations.

Deep marshes are dominated by *emergent* plants, species that are rooted in soil but grow mainly above the surface of the water. The most widely recognized marsh plants are the cattails. Two species, narrowleaf and broadleaf cattails, indicate relatively greater alkalinity and acidity respectively; narrowleaf cattail is also more salt-tolerant and therefore dominates near the coast. Cattail stands may be nearly uniform or richly interspersed with other wetland species such as arrow arum, sweetflag, irises, wetland umbellifers, and bur-reeds. The fact that the leaves and stems of these emergent species persist year-round distinguishes marshes from deeper-water communities, where emergent plants disappear below the surface after the growing season. *Shallow marshes* are typically dominated by grasses, sedges, and rushes.

Wet meadows, which may be flooded for only a brief period annually, are classified by some authorities as the driest form of marsh. They are also dominated by grasses, sedges, and rushes but typically contain more wildflowers than wetter communities and in New England are often artificially maintained by grazing or mowing.

Another type of marsh, often distinguished as a separate community, is the *tidal freshwater marsh,* which occurs in a zone near the mouth of a river between coastal salt marshes and nontidal freshwater marshes upstream; it is a relatively limited habitat in New England with only about a thousand acres total in Massachusetts, mainly along the North and Merrimack rivers. It has distinctive "field marks"—mainly related to the pronounced tidal fluctuation in water level. It can be recognized by the presence of certain dominant plant species such as soft-stemmed bulrush and sweetflag and contains a number of rare plant species— Eaton's beggar-ticks (ST), river bulrush (SSC), Parker's pipewort (SE)—that occur nowhere else.

Freshwater marshes may cover hundreds of acres, for example, in places where a slow-flowing river spreads out into a shallow basin. More often in New England, smaller marshes form at the shallow edges of ponds and lakes or sluggish backwaters of rivers and streams. A marsh may also be a stage in the life cycle of a pond: Under stable conditions, the pond eventually fills with enough dead vegetation to support a deep marsh, then a shallow marsh, then a shrub swamp, then a red maple swamp, and so on.

While wetland classification systems are useful for drawing distinctions and writing laws, a little field experience rapidly reveals a general pattern of intergradation not only between different types of marsh but between marshes and other types of wetlands such as swamps (see also BOG, LAKES AND PONDS, CALCAREOUS FEN, SALT MARSH).

Marshes are mysterious, haunting, somewhat forbidding places. In broad expanses they can instill the same mood of dank desolation as an English moor, a sensation exacerbated when cold mists rise in the predawn chill or a rising bubble releases a breath of fetid marsh gas (methane). While it is possible to tread carefully through some marshes in the dry season, one should always expect to emerge wet and muddy at least to the knees, and there is always a nagging anxiety that a wrong step will draw you down forever into the marsh wife's brewery. For the naturalist there is the additional mystery and fascination of the odd and secretive birds that live in marshes. If you have never stood by a marsh before dawn during the breeding season and listened to the wails, moans, cackles, and lunatic laughter with which rails, bitterns, gallinules, and other marsh birds express their territorial jealousies and sexual longings, you

Painted turtle

have missed one of nature's strangest performances. Alas, the experience is becoming all too rare as these wetlands are fragmented and degraded and their bird populations decline precipitously.

INDICATOR SPECIES

Wet meadow species are not included. ∞ GRASSES, SEDGES, AND RUSHES: Wild rice (especially tidal freshwater marshes), common reed (disturbed sites), reed canary-grass (SM, X; often forming dense stands), rice cutgrass (SM), blue joint (SM), rattlesnake grass (SM), woolgrass on common bullsedge, northern bullsedge (SM), softstem bullsedge, hardstem bullsedge, Small's spikesedge, tussock sedge (SM; may form a distinctive tussock marsh over large areas; sometimes distinguished as a separate sedge meadow community), marsh sedge, sallow sedge (SM), long-culmed sedge (SM), threeway sedge (SM), marsh or Canada rush (SM), soft rush (SM), short-tailed rush (SM). ∞ FERNS AND HORSETAILS: Marsh fern, royal fern, swamp horsetail. ∞ WILDFLOWERS: Broadleaf cattail (O; dominant in deep marshes), narrowleaf cattail (O; dominant in deep marshes and coastal and alkaline sites), sweetflag (calamus), great and lesser bur-reeds, common arrowhead, pickerel weed, arrow arum (O), larger blue flag, marsh bellflower, yellow loosestrife, purple loosestrife (X), marsh St. John's-wort (SM), water parsnip, bulb-bearing water hemlock (deadly poisonous), swamp milkweed (SM), boneset (SM), Joe Pye weed (SM), purple-stemmed aster. ∞ OLIGOCHAETE WORMS: Along with midges and mosquitoes, these worms are the most abundant macroinvertebrates in marshes. ∞

KEY
O=Obligate
SM=Shallow Marsh
SE=State Endangered
ST=State Threatened
SSC=State Special Concern
X=Exotic

MOLLUSKS: Flat-coiled gyraulus, modest gyraulus, Say's toothed planorbid, fingernail clam species. ∞ CRUSTACEANS: Mystic Valley amphipod (SSC). ∞ MAYFLIES: Small minnow mayflies, spiny crawler mayflies, diminutive squaregill mayfly. ∞ ARACHNIDS: Many species of water mites (order Acarina). ∞ DRAGONFLIES: No New England odonate species prefers cattail marshes to the exclusion of other habitats. Most spreadwing damselfly species favor marshy borders of ponds and lakes; mottled and Canada darners prefer marshy ponds; white-spangled skimmer, green jacket, and most meadowflies occur in marshes. Many other species occur in marsh-edged streams and ponds and in wet meadows. ∞ TRUE BUGS: Marsh treaders, many species of water boatmen. ∞ BEETLES: Marsh beetles, minute marsh-loving beetles, long-horned leaf beetles. ∞ BUTTERFLIES: Eyed brown, bronze copper (SM; rare; caterpillar feeds on water and curled dock, tussock, and other marsh sedges), least skipper (several wetland grasses and sedges), northern broken dash, mulberrywing (tussock sedge; adult nectars frequently on swamp milkweed), broad-winged skipper (common reed), black dash (tussock and other marsh sedges). ∞ TRUE FLIES: 80 percent of all insects emerging from marshes are chironomid (nonbiting) midges and mosquitoes. Many species of craneflies, phantom crane fly, soldier flies, long-legged flies, marsh flies. ∞ FISHES: Northern pike (X) and chain pickerel both spawn in marshes. Common carp (X), banded killifish, mud minnow (X). ∞ AMPHIBIANS: Most common species of frogs occur in shallow marshes and open areas of deep marshes (see LAKES AND PONDS). ∞ REPTILES: Snapping turtle, spotted turtle (SSC), painted turtle, Blanding's turtle (ST), northern water snake. ∞ BIRDS: Pied-billed grebe (ST), American bittern (O, ST), least bittern (O, ST). Other herons frequent marsh edges and openings. Green-winged teal, American black

Marsh wren

duck, mallard, northern pintail, blue-winged teal, northern shoveler, gadwall, king rail (O, ST), Virginia rail (O), sora (O), common moorhen (O, SSC), black tern (migrant), willow flycatcher, marsh wren, yellow warbler (frequents shrubby edges), common yellowthroat, swamp sparrow (O), red-winged blackbird (O). ∿ MAMMALS: Muskrats play a major role in shaping marsh structure by creating openings in dense vegetation. Star-nosed mole, beaver, mink.

DISTRIBUTION

Freshwater marshes occur worldwide wherever the appropriate aquatic conditions prevail. In North America there are great concentrations of freshwater marsh habitat in the northern prairie states, the remaining vestiges of the once extensive pothole marshes throughout the Midwest, the tule marshes of California, and the Florida Everglades, which are essentially a vast sawgrass marsh. Extensive cattail marshes are not common in southern

New England, though there are some notable exceptions in Massachusetts such as the Lynnfield Marsh, and extensive tidal freshwater marsh along the North River in Marshfield and in Essex County. Smaller marshes associated with ponds, lakes, and streams, and other wetland systems are nearly ubiquitous, with small examples to be found in most towns of the Commonwealth.

EXTENT IN MASSACHUSETTS

MacConnell (1975, see RESOURCES) gives the following figures for 1971–72: Wet meadows: 10,954 acres; shallow marsh: 23,286 acres; deep marsh: 16,200 acres; total: 50,440 or approximately 1 percent of the Commonwealth's total land area. Most of these sites are small, but a few comprise 200 or more acres.

CONSERVATION STATUS

During the seventeenth, eighteenth, and early nineteenth centuries, many acres of marshland were drained and converted to cropland. Further destruction of marshland for industrial, commercial, and residential use continued a process that abated only with the passage of wetlands protection laws in 1963 and 1967. The present Massachusetts Wetlands Protection Act combined these laws in 1972. This legislation was amended in 1986 to provide additional protection for wetland wildlife. However, recent analysis by the Massachusetts Department of Environmental Protection indicates that these wildlife protection provisions are largely ineffective. Overall, marshes and other wetlands continue to be lost and degraded, in part as a result of various kinds of projects that are still permitted under the Wetlands Protection Act. Even

Opposite: Broadleaf cattail marsh

small projects result in fragmentation and cumulative degradation, resulting in a significant loss of biodiversity.

Public apathy permits such losses. As our cultural traditions have moved away from the land and become more urban and indoor-oriented, the perception of marshes as inhospitable, smelly, mosquito-infested wastelands has grown, with predictable results. In addition to outright destruction, many marshes are subject to degradation from pollution of their source waters. Some of these marshes appear healthy superficially, but have lost the populations of marsh birds and other elements of biological diversity they once held.

Another major threat to freshwater marshland is invasion by aggressive alien and weedy plant species, especially purple loosestrife and common reed (Phragmites). Introduced in the early 1800s both deliberately as a valuable medicinal herb (for diarrhea, bleeding, sores, ulcers, and wind) and accidentally in ship's ballast, purple loosestrife spreads rapidly via waterborne seeds and prodigious productivity. It was well established along the New England coast by the 1830s and its spread inland was aided by the construction of canals and other waterways. Phragmites is a native grass that occurs naturally in brackish estuaries and salt marsh edges, but it is also stimulated by disturbances such as road-salt runoff and burning to invade freshwater marshes. This disturbance-prone form may be genetically distinct. Both loosestrife and Phragmites compete aggressively with cattails and other native marsh plants, drying out the habitat and eventually making it unsuitable for many native plants and marsh-breeding animals. Once established, these plants are virtually impossible to eradicate except by altering the hydrology or resorting to repeated applications of herbicides. The undesirability of these alternatives has fueled a lively public controversy over the dense stands of Phragmites that have invaded Boston's Back Bay Fens. Recently some states have permitted the introduction of several Old World beetle species that feed on loosestrife, in the hope of controlling the weed without resorting to toxic chemicals.

PLACES TO VISIT

Lynnfield Marsh, Towns of Lynnfield and Wakefield.

Great Meadows National Wildlife Refuge, Concord. U.S. Fish and Wildlife Service.

Parker River National Wildlife Refuge (access from Newburyport). U.S. Fish and Wildlife Service.

Freetown State Forest, Freetown. Massachusetts Department of Environmental Management.

North River Marshes (access, especially by canoe, from Hanover, Pembroke, and Norwell).

Great Wenham "Swamp," Ipswich River Wildlife Sanctuary, Topsfield. Massachusetts Audubon Society.

FURTHER READING

The Ecology of Tidal Freshwater Marshes of the United States East Coast: A Community Profile, by William Odum et al. Washington, D.C.: U.S. Fish and Wildlife Service, 1984.

Freshwater Marshes, by Milton Weller. Minneapolis: University of Minnesota Press, 1987 (2nd ed.).

Wetlands, by William Niering. New York: Knopf, 1985. An Audubon Society Nature Guide.

Great blue heron

Lakes and Ponds

In the puddle or pond, in the city reservoir, ditch or Atlantic Ocean, the rotifers still spin and munch, the daphnia still filter and are filtered, and the copepods swarm hanging with clusters of eggs. These are real creatures with real organs leading real lives, one by one. I can't pretend they're not there. If I have a life, sense, energy, will, so does a rotifer.

ANNIE DILLARD
Pilgrim at Tinker Creek

Bunker Meadow, Ipswich River Wildlife Sanctuary, Topsfield

Lakes and ponds are still, or *lentic,* bodies of fresh water created both naturally and by people. In New England many of the pits and scratches now filled with water were made by glaciers. One of our most common pond types lies in kettleholes, formed where blocks of glacial ice were embedded in the soils of outwash plains. If the pot-shaped depression remaining after the ice melted reached below the water table, it became a pond filled with groundwater. When a river straightens itself out by cutting off one of its laziest meanders, the cast-off bend becomes an oxbow pond. A long deep lake may be created when the earth's tectonic plates shift and open a great rift that fills with water. And many still bodies are simply depressions underlain with impermeable rock or clay (or plastic).

Beavers create ponds by constructing dams of branches and mud across streams, flooding the land behind. Having returned during the last century from the brink of extinction, they are busily creating extensive areas of new aquatic habitat and biodiversity, to the delight of naturalists and the annoyance of some land owners.

Most of the ponds in Massachusetts were made, or "enhanced," by another industrious mammal using shovels and backhoes. The very word *pond* (originally *pounde*) meant an artificial body of water or empoundment created to provide water for various human, especially agricultural, uses.

125

Thousands of millponds were created in the Commonwealth in its agricultural heyday by damming streams and rivers to provide power for grinding grain. With their constant water flow and higher oxygen content millponds are often quite different biologically from ponds that are simply dug into the water table.

Like frogs and toads, ponds and lakes beg to be defined in terms of each other. Most people would probably say that the principal difference between a pond and a lake is size. There is some truth in this: No one would think of describing Lake Superior as a pond, nor would one glorify the local pout pond as a lake. However, to a limnologist (someone who studies bodies of fresh water), the critical factor is depth. By definition ponds are shallow enough that light can reach the entire bottom, allowing plants to grow from shore to shore; they do not "stratify" (possess different temperature zones); and they may freeze to the bottom in winter. Lakes are deeper and therefore darker, with a portion of the bottom unable to support photosynthesis. By this definition there are some postage-stamp lakes (such as old quarries) and

some ponds the size of small seas. In Massachusetts many of the bodies we call ponds are technically small lakes.

Ecologically, lakes are more complex than ponds. In ponds the food web is based on the growth of rooted plants and algae. In lakes, where light does not reach much of the bottom, much of the energy production is accomplished by phytoplankton—microscopic green and blue-green algae and diatoms that drift near the surface of the water.

In a typical healthy pond in midsummer the water is dark green, with algae growing on the bottom and in the water column. Floating-leaved plants such as duckweed and spatterdock cover much of the surface. Snails, insects, crustaceans, worms, and tadpoles graze on the algae and plant material and on the decaying plant debris that makes up most of the pond bottom. Zooplankton feed on fine particles of algae and detritus as well as bacteria. Warm-water fish such as golden shiners, yellow perch, and sunfish prey on the zooplankton, while bottom-feeding bullheads and suckers scavenge in detritus and sediment. Turtles,

Common loons

snakes, and frogs forage and bask at the surface, and top predators such as largemouth bass gobble smaller vertebrate prey and large invertebrates.

In lakes with warm shallows supporting rooted plants along the shore, life in this "littoral zone" is much like that in a pond. In large lakes with heavy wave action, typical stream species adapted to strong current and high oxygen levels often occur along the shore. In the deeper water of lakes zooplankton such as water fleas, copepods, and rotifers are more abundant and diverse than in ponds, and there is likely to be a greater variety of native fish species. In the dark, cold profundal zone of a lake, the bottom consists of fine, mucky sediment, the remains of myriad planktonic organisms. Feeding in this sediment are thousands of oligochaete worms, bloodworms (larvae of chironomid midges), and phantom midge larvae.

Ponds typically change gradually over time. Any pond begins as a relatively sterile quantity of water in a basin of soil or rock. A few pioneer plants and animals capable of surviving on sunlight and a few basic minerals enrich the system and attract a growing number of consumer organisms: mol-

lusks, crustaceans, insects, amphibians, and fish. As it matures, a pond may eventually become so rich in nutrients (eutrophication) that the oxygen in the water is depleted, suffocating many of the inhabitants. At the same time, vegetation emerging from the shallows as well as floating and submerged plants increase and decompose, slowly filling in the pond. In time cattails and other emergents may be able to grow across the entire surface, creating a marsh, which in turn may be invaded by shrubs and trees and become a red maple swamp. Such a process may take decades or eons in ponds, depending on any number of variables, and this aging process does not necessarily occur at all in lakes.

Lakes undergo distinct seasonal changes. As one might expect, the water nearest the surface is much warmer during the summer than that nearest the bottom. The surface water also contains more oxygen and is therefore lighter—less dense. There is a distinct boundary—called the "thermocline"—a zone of rapidly descending temperature between the bottom and surface layers; many organisms can survive only on one side or the other of this frontier. As the surface cools in the fall, the temperature of the two layers equalizes and many animals can circulate throughout the water column. Once the surface temperature drops below 4°C (39°F) the pond "stratifies" again for the winter, with the warmest water on the bottom. This allows the surface to freeze while the warmer depths remain open and available as aquatic habitat, as anyone who has watched turtles swimming under the ice of their skating pond (or rather skating *lake!*) can testify. As the lake's surface warms again in the spring, another "turnover" begins. In ponds, by contrast, the shallow water is mixed constantly by the wind during the warmer months and can freeze to the bottom in winter.

Lake and pond communities can be divided into a number of habitat layers and zones from the air above the pond to the soil of the pond bottom and from the deepest area of the pond landward. Characteristic plants and animals may be restricted to one of these habitats or range across two or all of them. Flying organisms such as swallows or adult dragonflies interact with the pond from above. Whirligig beetles and floating plants live on the surface of the water. Fish swim through the middle layers, some preferring warmer waters near the surface and others thriving only in the cool depths. Many insect larvae are strictly bottom dwellers, crawling over or burrowing into the mud

until they are ready to assume their adult form and move to a different part of the community.

Shifting from the vertical to the horizontal plane, a vegetational progression appears: the open center of a lake with no plant growth; then a zone of submerged plants such as milfoils rooted in the bottom; next a zone of floating species such as water lilies; followed by an area of nonpersistent emergent plants such as pickerel weed, which die back in the winter; and finally a zone of emergents in shallower water such as cattail, which persist year-round. This zone might be followed by areas of shrub and forested swamp before reaching upland habitats. Seen from above in an idealized lake, these zones would look like concentric rings corresponding to water depth, each ring with its own constituent species. By the strictest definition, true pond life encompasses the zones of floating species and of nonpersistent emergent plants (pickerel weed); the zone of persistent emergents such as cattails technically constitutes a marsh. Of course the number of zones present varies from pond to pond (as well as from limnologist to limnologist).

The lake/pond is arguably the natural community that offers the greatest diversity of pure fun. Whether an abandoned farm pond or a cool kettlehole pool in the wilderness, a biologically well-developed body of this kind can probably reward infinite curiosity and scrutiny at a relaxed pace in addition to serving as a swimming and fishing hole. A dip net, a jar, a magnifying glass, a bathing suit, a fishing pole, and a cold box, applied to a small lake of varying depth on a bright summer's day has it all over any amusement park.

Snapping turtle

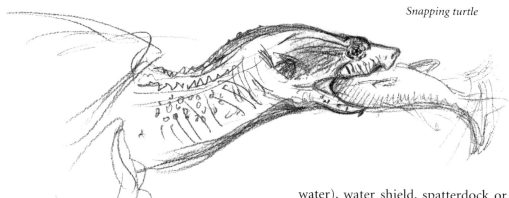

Snapping turtle

INDICATOR SPECIES

Lake species, that is, those preferring deeper colder water, are marked (L); the remaining species may be found in ponds and in shallower margins of lakes where rooted plants occur. For trees, shrubs, and wetland wildflowers commonly associated with ponds, see appropriate sections of FRESHWATER MARSH and RED MAPLE SWAMP. ∾ SUBMERGED PLANTS: Some submerged plants may have emergent flower stalks. Water moss, 2 native species of naiad and 2 aliens, horned pondweed, water stargrass, pinnate-leaved water-milfoil (SSC) and 5 other native species and 2 aliens, coontail, water starwort. ∾ FLOATING PLANTS: About 25 species of *Potamogeton* pondweed may be rooted in bottom or have roots suspended in water. Mosquito fern, water shamrock, 5 species of duckweed, 4 species of water meal, 12 species of bladderworts, water smartweed (several other members of the genus *Polygonum* occur in shallow

KEY

A=Anadromous
C=Catadromous
L=Lake species
FE=Federally Endangered
FT=Federally Threatened
ST=State Threatened
SSC=State Special Concern
X=Exotic

water), water shield, spatterdock or yellow pond lily, southern pond lily, small pond lily, floating heart, sweet-scented water lily, water marigold, 2 species of yellow water buttercup, 2 species of white water buttercup, water purslane, waterwort. ∾ EMERGENT PLANTS: 5 species of quillworts, 7 species of bur-reeds, large and lesser water-plantains, golden club (ST), arrow arum, 5 species of arrowheads, common pipewort, pickerel weed, spikesedges (several species grow as emergents), soft and Canada rushes, water chestnut (X; pest), mermaid-weed, mare's tail, American lotus, water willow (shrub).

The species of pond invertebrates are so numerous that generally only key groups and the most common or most characteristic species are listed. ∾ WORMS: New England medicinal leech (SSC), other common leech species, many species of oligochaete worms. ∾ MOLLUSKS: Lake fingernail clam (L; formerly widespread in eastern Massachusetts but now uncommon), pond fingernail clam, triangular pea clam, common spire snail, three-keeled valve snail, brown mystery snail, eastern tadpole snail, irregular and modest gyraulus, keeled promenetus, Say's toothed planorbid, large eastern ramshorn snail, pointed sand-shell mussel, eastern elliptio, heavy-toothed wedge mussel, squawfoot (SSC), eastern lamp mussel, delicate lamp mussel. ∾ MAYFLIES: Small minnow mayflies, primitive minnow mayflies, burrowing mayflies, square-gilled mayflies, spiny crawlers. ∾ DRAGONFLIES: Spreadwing damselflies (all 9 New England species inhabit ponds and lakes,

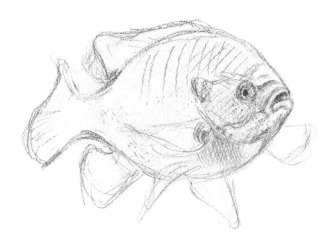

Bluegill

mosquitoes. ❧ **ARACHNIDS:** 3 species of fishing spiders, red water mites. ❧ **CRUSTACEANS:** Seed shrimps, isopods, water fleas, copepods, amphipods (ubiquitous in aquatic plant beds), crayfish. ❧ **FISHES:** American eel (C), alewife (A), goldfish (X), lake chub (L, SE), common carp (X), golden shiner, bridle shiner, mimic shiner (L, X), bluntnose minnow (L, X; commonest minnow in the Quabbin Reservoir), fathead minnow (X), rudd (X), fallfish (L), longnose sucker (L, SSC), white sucker, creek chubsucker (L), white catfish (X), yellow bullhead (X), brown bullhead, channel catfish (L, X), tadpole madtom (X), redfin pickerel, northern pike (L, X), tiger muskellunge (L, X; a sterile hybrid between northern pike and muskellunge), chain pickerel, central mudminnow (X), rainbow smelt (L), rainbow trout (L, X), lake trout (L, X), trout perch (L; extirpated), banded killifish, ninespine stickleback, slimy sculpin, white perch, rock bass (L, X), banded sunfish, redbreast sunfish, green sunfish (X), pumpkinseed, bluegill (X), smallmouth bass (X), largemouth bass (X), white crappie (X), black crappie (X), swamp darter, yellow perch, walleye (X). ❧ **AMPHIBIANS:** Red-spotted newt, American toad, bullfrog, green frog, northern leopard frog, pickerel frog. ❧ **REPTILES:** Common snapping turtle, musk turtle (stinkpot), spotted turtle (SSC), red-eared slider (X), Plymouth redbelly turtle (FE; southeastern coastal plain), eastern painted turtle, Blanding's turtle (ST), northern water snake. ❧ **BIRDS:** Common loon (L, SSC), pied-billed grebe (ST), double-crested cormorant (L), great blue heron, green heron, black-crowned night heron, mute swan (X). Twenty duck species occur regularly on ponds, but only the mallard and black duck commonly breed in this habitat: canvasback, redhead, and ring-necked ducks, greater and lesser scaups, common merganser, and ruddy duck all show an affinity for lakes. American coot (L), bald eagle (L, FT), osprey, common moorhen (SSC),

especially ones with weedy edges), blue-fronted dancer, dusky dancer, bluets (most of our 21 species prefer still waters; familiar bluet is characteristic of larger lakes), lilypad forktail, eastern forktail, common sanddragon, beaverpond clubtail (SSC), lancet clubtail, lilypad clubtail, dusky clubtail, unicorn clubtail, cobra clubtail (SSC), common green darner, lake darner (L; most darners in the genus *Aeshna* frequent lakes and ponds), ocellated darner (L, SSC); swift river cruiser (L), stygian shadowfly (L; very rare), petite emerald, racquet-tailed emerald, coppery emerald (SE), American emerald, beaverpond bottletail, common baskettail, prince baskettail (L), spiny baskettail (L), eastern amberwing, all 4 pennant species, all 12 skimmer species, eastern pondhawk, blue dasher, all 10 meadowfly species, all 5 whiteface species, violet-masked glider. ❧ **TRUE BUGS:** Backswimmers such as the common grouse-winged backswimmer, giant water bug, and water boatmen (species in 4 or more genera); water scorpions, water striders, water treaders, pond bugs, creeping water bugs. ❧ **BEETLES:** Predaceous diving beetles, various crawling water beetles such as the twelve-spotted, whirligig beetles. ❧ **CADDISFLIES:** Swimming caddisflies, micro-caddisflies, large cigar-case makers, northern case makers (3 species), 2 log-cabin caddisflies, hooded flange-case makers (*Molanna* species). ❧ **TRUE FLIES:** Phantom midges, chironomid midges, craneflies,

Bluegill

spotted sandpiper, ring-billed gull (L); other gulls commonly bathe and loaf in ponds and lakes. Belted kingfisher, swallows (all 5 species commonly feed over ponds), eastern kingbird, yellow warbler, and common yellowthroat frequently nest in shrubby vegetation near ponds. ❧ MAMMALS: Water shrew (SSC), bats (most of our species feed over ponds at least occasionally), beaver, muskrat, raccoon, otter, mink.

DISTRIBUTION

Ponds are common communities throughout the planet from the Arctic and Subantarctic to the tropics. New England and Massachusetts have ponds in extraordinary abundance for three reasons: the earth-gouging nature of glaciation, the presence of beavers, and the coming of the European pond makers in the seventeenth century.

EXTENT IN MASSACHUSETTS

According to the Massachusetts Division of Water Pollution Control (1988) there are 2,859 lakes and ponds in Massachusetts, covering a total of 150,340 acres; 43 percent of them are less than 10 acres in area. Many of these water bodies are artificial or "enhanced," the percentage varying widely from region to region: western Massachusetts, 70 percent; central Massachusetts, 93 percent; eastern Massachusetts, 51 percent; Cape Cod, 20 percent.

CONSERVATION STATUS

Lakes and ponds seem inordinately susceptible to a wide range of ecological maladies. The major ones are summarized below.

Cultural Eutrophication

Eutrophication refers to the enrichment of a body of water by nutrients, mainly nitrogen and phosphorus. Most ponds eutrophy naturally as they age as a result of increasing plant growth and decomposition. Cultural eutrophication is the *artificial*

enrichment of lakes and ponds due to leaking septic systems or stormwater runoff, which delivers lawn fertilizers and organic wastes from streets and other paved surfaces into water bodies. This typically leads to lush blooms of aquatic weeds and algae, often followed by the destruction of the biological, recreational, and aesthetic benefits that ponds and lakes provide. Unfortunately, rather than addressing the source of the pollution, many people attempt to treat the symptoms of eutrophication with herbicides, weed rakes, dredging, draw-downs (the draining of ponds to kill plants by drying and freezing), and the introduction of exotic fishes (illegal in Massachusetts). In addition to being ineffective, such stopgap measures may further degrade the natural biological diversity of the system. The only long-term solution is to protect the watershed: maintain vegetated buffers around the shoreline, prevent septic system discharges, and keep nutrient levels in runoff low.

Aquatic Weeds

A related "problem" involves *naturally* eutrophic ponds and lakes and human perception of what constitutes an ideal body of still water. As described earlier, a healthy pond ecosystem grows thick with algae and native weeds by midsummer. Indeed these plants are the basis of the food web in such habitats. But to many people they are just an unattractive nuisance that interferes with swimming and boating. It is depressing to consider how much money is spent by pond-shore residents on herbicides, motorized weed harvesters, and draw-downs in a vain attempt to turn a splendid natural pond filled with wildlife into a temporary facsimile of a sterile alpine tarn. The solution here must be education. A once-clear lake that has been choked to death by alien weeds as a result of human carelessness amounts to fouling one's own nest, a problem to be solved. But a naturally eutrophic pond crammed with water lilies, turtles, and dragonflies is a biological paradise to be treasured and protected.

Exotic Species

The introduction of aggressive alien organisms plagues many of our ponds and lakes. Such species can take over an ecosystem, replacing native species and destroying biological and recreational values. Boaters transport exotic water weeds, such as Eurasian water milfoil, which may proliferate rapidly. Unused exotic bait species, including the rusty crayfish and several species of "minnows," are often dumped by fishermen, with the result that very few of our ponds now contain wholly native fish faunas. Furthermore, for many years a primary goal of fish and wildlife agencies was to stock ponds and lakes with sport fish, including alien species; the wisdom of this practice, which continues today, is debatable.

Birds

Another problem of our own making is the pollution of ponds and lakes by large concentrations of certain species of birds, especially in urban and suburban areas. Populations of gulls, and feral waterfowl such as Canada geese, mute swans, and mallards have been inflated to unhealthy proportions by human activities such as dumping, commercial fishing practices, the release of exotic species, and feeding of birds at "duck ponds." Enough birds defecating in a pond or lake can cause rapid eutrophication and eventually the suffocation of the ecosystem. Such avian cesspools are also unhealthy for birds and people.

Acidification

Most Massachusetts soils are relatively acidic by nature, and the chemistry of our lakes and ponds reflects this. Small, isolated basins with little flow tend to become highly acidic and nutrient-poor (see BOG and CALCAREOUS FEN). Larger bodies fed by groundwater and subject to flushing are generally well buffered and not currently vulnerable to acid deposition. In certain parts of the state, however, lakes and ponds have become depleted of the dissolved salts that can neutralize acids. When

Pickerel frog

this happens, plant growth and algal photosynthesis decline and the richness of the fauna decreases dramatically. Recent improvements in air pollution regulations have led to decreases in the amount of acid precipitation falling in Massachusetts. If such measures stay in place, the Commonwealth may be spared the widespread mortality of lakes and ponds (and forests) witnessed in other parts of the world.

PLACES TO VISIT

Few towns or conservation areas are without one or more accessible ponds or lakes, and even the poorest of these is likely to be popping with aquatic life in summer. All lakes or ponds that were 10 acres or larger in their natural state (before damming) are defined as "great ponds" and belong to the Commonwealth, with use guaranteed to its citizens. Unfortunately—as with the seashore—access to such ponds over private property is *not* guaranteed. Massachusetts Audubon wildlife sanctuaries with first-rate ponds are *Ipswich River,* Topsfield; *Habitat,* Belmont; *Stony Brook,* Norfolk; *Daniel Webster,* Marshfield; *North Hill Marsh,* Duxbury; *Wellfleet Bay,* Wellfleet; *Felix Neck,* Martha's Vineyard; *Wachusett Meadow,* Princeton; *Laughing Brook,* Hampden; and *Canoe Meadows,* Pittsfield. For a listing of public ponds and lakes in Massachusetts, write the Massachusetts Division of Fisheries and Wildlife (see RESOURCES, p. 208).

Though manmade, the Quabbin Reservoir is indisputably Massachusetts' "great lake," with a surface area of 39.4 square miles and breeding populations of characteristic wilderness lake species such as the common loon and bald eagle. It is owned and managed by the Metropolitan District Commission; public access to the Quabbin Reservation is in Belchertown. Wachusett Reservoir and other public water supplies are also good examples of deep lakes with heavy wave action and contain species adapted to these conditions.

Two especially visitor-friendly ponds in state parks are:

Berry Pond, Harold Parker State Forest, North Andover. Massachusetts Department of Environmental Management (DEM).

Dunn Pond State Park, Gardner, DEM.

FURTHER READING

A great many guides and other books about pond life have been published over the years. The following current or classic titles are a small selection.

Aquatic Entomology, by W. Patrick McCafferty. Boston: Science Books International, 1981.

The New Field Book of Freshwater Life, by Elsie B. Klots. New York: G.P. Putnam's Sons, 1966.

"Policy on Lake and Pond Management," by Louis J. Wagner. Lincoln, Mass.: Massachusetts Audubon Society, 1993.

Pond Life, by George K. Reid. New York: Golden Press, 1987.

Pond map booklets, published by the Massachusetts Division of Fisheries and Wildlife. Six regional collections of pond maps providing bottom contours, public access sites, and general fisheries information on all the major ponds in Massachusetts.

Pond Watcher's Guide to Ponds and Vernal Pools in Eastern North America. Lincoln, Mass.: Massachusetts Audubon Society, 1995.

Quabbin: The Accidental Wilderness, by Thomas Conuel. Amherst, Mass.: University of Massachusetts Press, rev. ed. 1989.

Watchers at the Pond, by Franklin Russell. New York: Time-Life Books, 1961.

Wetlands, by William Niering. New York: Knopf, 1985. An Audubon Society Nature Guide.

Green heron

Rivers and Streams

Never in his life had he seen a river before—this sleek, sinuous, full-bodied animal, chasing and chuckling, gripping things with a gurgle and leaving them with a laugh, to fling itself on fresh playmates that shook themselves free, and were caught and held again. All was a-shake and a-shiver—glints and gleams and sparkles, rustle and swirl, chatter and bubble....He sat on the bank, while the river still chattered on to him, a babbling procession of the best stories in the world, sent from the heart of the earth to be told at last to the insatiable sea.

KENNETH GRAHAME
The Wind in the Willows

River, stream, creek, spring—these terms have no precise definitions, but they do present distinctive pictures to the mind's eye, suggesting—correctly—several natural communities rather than one. Yet a single factor serves to unite them and at the same time distinguish them from most other freshwater bodies: current.

All of the factors that influence life in other aquatic systems—light, depth, turbidity, temperature, chemistry—also apply to rivers and streams. The big difference is the force of gravity. The unique challenge (and opportunity) of river life is that the community is in constant, one-way, sometimes rapid motion because it is moving downhill, carrying whatever excess water, eroded sediment, and organic matter runs off the land or drops into the water as it moves. As the river channel widens or narrows, or the slope of the descent alters, the speed of the water changes accordingly. The height and breadth of the river is subject to drastic change according to the rate of precipitation. The constant motion ensures that the water is continually recharged with oxygen. The waters are constantly mixing, with little or none of the stratification characteristic of lakes. And the bottom is swept and scoured without pause of anything attempting to be sedentary by the current and its cargo. In short, a river is a highly dynamic, open-ended system where the only constant is change.

Every river species has evolved mechanisms to utilize, counter, or at least survive the characteristics of current, and observing these different ways of going with or against the flow gives us the best insights into the nature of life in lotic, or running, waters. A few examples must suffice. Many riparian plants have long, complex root systems that anchor them to the soggy, fluctuating banks. Other plants, including the showy but pernicious

Opposite: Money Brook, cold water stream in "the Hopper," Mt. Greylock

purple loosestrife, use the streams to extend their range by dispersing floating seeds or other parts. Many stream invertebrates tend to have flat and/or contoured body shapes, allowing fast-moving water to flow over or around them with minimal drag, and are adorned with all manner of hooks, life lines, and suckers to keep them from being flushed out of their preferred ecological niches. The larvae of a common family of caddisflies, the net-spinners, make catch-nets to trap bits of detritus; different species apparently use different mesh sizes to accommodate the speed of the current and their food preferences. River fishes are neurologically equipped to maintain a constant position in relation to the current without (so to speak) thinking about it—a trick called rheotaxis. Many river-haunting bird species—for example, spotted sandpiper and Louisiana waterthrush—seem to have evolved characteristic plumage and movements that blend cryptically with the play of light on moving water. And the streamlined, muscular, web-footed otter has the option of simply climbing ashore when it gets tired of using its river skills.

While it is slightly exhausting to think about riverine organisms in eternal Sisyphean struggle against the current, we should bear in mind that these species' identities have evolved in concert with their restless medium, and many would not survive long in any other. The nymphs of stream dragonflies would rapidly go belly up in relatively airless pond water, though some such species also thrive in the highly oxygenated shallows of large lakes with heavy wave action.

It is important to understand the unifying nature of the gravity-driven current, which creates a continuum, from mountainside seep to a great river's merging with the sea. At the same time we should not lose sight of the variety of conditions that exist throughout the broad descending network. For example, the primary food source in small headwater streams comes mainly from the shredding by microorganisms of vegetation that falls in from overhanging trees and bankside vegetation. As you move down the watershed, water flow increases and the stream channels become more exposed to sunlight, so photosynthesis by instream algae and plants can take over as the principal energy source. Here, invertebrate grazers and gatherers replace the shredders that prevailed above. By the time you arrive in the lower reaches of the watershed, the water is likely to be relatively deep and turbid, and photosynthesis again becomes less significant. Now, in the river's main stream, transported detritus and nutrients as well as woody debris and periodic inputs from the floodplain serve as the basis for much of the food web.

People tend to perceive rivers as moving water, but to riverine plants and animals the characteristics of the bottom are at least as important. This riverbed varies greatly along the course of a stream system according to the force of the flow and type of land through which the water flows. Erosion scours out deepwater pools along bends or in eddies, deposits rocky spoils to form riffles, and, where the current is gentler, lays down finer sands and muds. These and other types of river architecture provide habitats for particular groups of organisms. In general, river invertebrates tend to be burrowers and soggy-wood specialists, whereas stream invertebrates operate more in the current and take advantage of the passing food supply.

Rivers have a distinct seasonality related to flow and temperature, and as with other communities, the river seasons are characterized by certain life forms. Spring brings floods, but also the migration and spawning of most river fishes and swarms of mayflies and other aquatic insects. Mid-June to early July is the peak period for the riverine dragonflies called clubtails. The late-summer and early-fall season typically sees the lowest water levels, opening river flats for migratory shorebirds and providing maximum habitat for species that, like the puritan tiger beetle, favor sandy river islands. The largest, fastest rivers often remain partially free of ice in winter, attracting ducks, raptors, kingfishers, and other aquatic predators. Even under the ice, many river life forms remain active despite low temperatures. Indeed, many stream invertebrates are most active in the winter, when they feed actively on fallen autumn leaves.

In the not-so-distant past, before cars, trains, and airplanes, rivers were our principal transportation arteries. Native Americans settled extensively along rivers, and these liquid highways provided the route by which Europeans explored and colonized the New World. In the early colonial period rivers were an essential means of transporting lumber, a mainstay of the export economy, to coastal ports. And in the early years the river fishery was almost as bountiful a resource as the coastal one. Rivers and streams were also crucial in the next stage of development, the Industrial Revolution. It is no exaggeration to say that the mills and factories responsible for the first "Massachusetts miracle" in the nineteenth century could not have functioned without the rivers and streams on which they invariably were situated. Worcester,

Lowell, Lawrence, Fitchburg, Springfield, and many other Massachusetts cities got their start and grew as river towns. Today, perhaps the most important benefit of rivers is drinking water, 66 percent of which comes from surface waters, mostly from reservoirs created by damming portions of the river system. Of course it is also the rivers that deliver water to most of our freshwater wetlands and regulate the hydrologic regimes in the wetlands and in standing bodies of surface water. The surface waters and wetlands in turn are among our greatest reservoirs of biological diversity.

There is no better way to feel a part of nature than to canoe at a leisurely pace down a river, especially at a time or place which other canoers have foolishly eschewed. The water below makes you feel more enveloped than you ever do hiking on a path. The flow exerts a certain control to which you can readily surrender without much anxiety. You can ogle the river flora at close range without sinking into littoral ooze. And for some reason, wildlife does not perceive paddlers to be as dangerous as people on foot. Dragonflies alight on prow and knee. Herons almost within reach respond with a nonchalance bordering on disdain, as to a floating log. Mink and otter sometimes go apoplectic and act as if they might leap aboard. And no other natural byway can inspire the same level of anticipation—of encountering a perched eagle, an incomparable vista, an unmapped dam—as rounding a river bend.

Spring salamander

White-tailed skimmer

Species preferring cold streams are often treated as constituting a distinct community and are identified in the list below as CS. Many of the plants and animals listed show some preference for rivers but are not restricted to the community. Excluded here are estuarine, marsh, and floodplain forest species and those that inhabit a wide range of aquatic habitats. ❧ **TREES, SHRUBS, AND VINES:** See FLOODPLAIN FOREST. ❧ **GRASSES AND SEDGES:** Wild rice, riverbank wild-rye, river bullsedge, spreading bullsedge, twisted sedge. ❧ **OTHER PLANTS:** 6 species of pondweeds (genus *Potamogeton*), including Hill's pondweed (SSC), favor slow-moving streams as well as ponds. Several species of arrowhead *(Sagittaria)* occur in river mud. See also FRESHWATER MARSH, LAKES AND PONDS, and FLOODPLAIN FOREST. ❧ **MOLLUSKS:** Striated fingernail clam, common spire snail, eastern elliptio, dwarf wedge mussel (FE), heavy-toothed wedge mussel, swollen wedge mussel (SE), eastern floater, alewife floater, squawfoot (SSC), pointed sandshell (coastal streams and Connecticut River Basin), yellow lamp mussel (SE; now only in Connecticut River), eastern lamp mussel, eastern river pearly mussel (CS). ❧ **CRUSTACEANS:** 9 species of crayfish, including 3 natives, 5 exotics, and 1 questionable; the Appalachian brook crayfish is CS and SSC. ❧ **MAYFLIES** (MAINLY COLD-STREAM SPECIES): Spiny crawler mayflies, including Hendrikson's mayfly, hacklegill, little stout crawlers, burrowing

---

KEY
A=Anadromous
C=Catadromous
CS=Species preferring cold-stream habitat
FE=Federally Endangered
SE=State Endangered
ST=State Threatened
SSC=State Special Concern
X=Exotic

mayflies, including green drake, brown drake, and yellow drake, giant Michigan mayfly, flatheaded mayflies, including quill Gordon, pale evening dun, March brown mayfly and other flatheaded mayflies, pronggill, black quill, willowfly (large rivers), and several species all known to anglers as light cahill. ❧ DOBSONFLIES: Horned dobsonfly. ❧ DRAGONFLIES: Ebony jewelwing, river jewelwing, sparkling jewelwing (very rare; mainly southeastern Massachusetts), superb jewelwing (CS), American rubyspot, several spreadwing species (slow streams with weedy banks), blue-fronted dancer, powdered dancer, dusky dancer, violet dancer, stream bluet, turquoise bluet, orange bluet (other bluet species sometimes frequent slow streams), delta-spotted spiketail, twin-spotted spiketail, arrow-headed spiketail (the spiketails favor shady forest streams), ocellated darner (CS, SSC), common sanddragon, dragonhunter, zorro clubtail, rusty snaketail, brook snaketail (CS, SSC), riffle snaketail (CS, ST), pygmy snaketail (very rare, Connecticut River; CS), Maine snaketail (CS), moustached clubtail, least clubtail, riverine clubtail (SE), rapids clubtail, harpoon clubtail (SE), ashy clubtail, arrow clubtail (ST), cobra clubtail (SSC), skillet clubtail (SSC), zebra (or Scudder's) clubtail (SE), brook clubtail (CS), black-shouldered spinyleg, stream cruiser, swift river cruiser, umber shadowfly (or twilight skimmer; SSC), stygian shadowfly (very rare), mocha emerald (SSC), forcipate emerald, ocellated emerald, Walsh's emerald, Williamson's emerald, Uhler's sunfly (CS). ❧ STONEFLIES (COLD-STREAM SPECIES): Four species of common stoneflies, little red stonefly, giant black stonefly, plus many smaller species of stoneflies.

Larger blue flag

CADDISFLIES (MAINLY COLD-STREAM SPECIES): Tortoise case-maker species, northern snailcase maker (makes a spiral case of sand that looks like a snail's shell), 16 to 18 species of scaly tube casemakers, 3 to 4 species of humpless casemakers, 12 to 14 species of free-living green caddisflies, net-spinner species (mostly in warmer water, often below old mill dams). BEETLES: 2 species of water pennies. FLIES: Blackflies are characteristic of cold streams. FISHES: American brook lamprey (ST), Atlantic sturgeon (A, SE), short-nosed sturgeon (A, FE), bowfin (X), American eel (C; migrates partly via rivers to spawning grounds in the Sargasso Sea), American shad (A), alewife (A), blueback herring (A), gizzard shad, lake chub (SE, CS, and also lakes), common carp (X), eastern silvery minnow (SSC; backwaters of big, slow-moving rivers), common shiner, golden shiner, bridle shiner, spottail shiner, mimic shiner (X), northern redbelly dace (SE, CS), bluntnose minnow (X), fathead minnow (X), blacknose dace (commonest stream minnow in Massachusetts), longnose dace (CS), rudd (X; lower Charles River), creek chub (CS), fallfish, longnose sucker (SSC, CS), white sucker, creek chubsucker, white catfish (X), yellow bullhead (X), brown bullhead, channel catfish (X), tadpole madtom (X), margined madtom (X), redfin pickerel, chain pickerel, rainbow smelt (A), Atlantic salmon (A), rainbow trout (X, CS), brown trout (X, CS), brook trout (CS), burbot (SSC, A), banded killifish, rainwater killifish (tidal freshwater), slimy sculpin (CS), rock bass (X), redbreast sunfish, bluegill (X), smallmouth bass (X), largemouth bass (X), black crappie (X), tesselated darter, yellow perch, walleye (X). AMPHIBIANS: Mudpuppy (X), bullfrog, spring salamander (SSC, CS), northern two-lined salamander (CS), northern dusky salamander (CS). REPTILES: Common snapping turtle, stinkpot (musk turtle), wood turtle (SSC), painted turtle, northern water snake. BIRDS: Double-crested cormorant, great blue heron, black-crowned night heron, ring-necked duck, hooded merganser, common merganser, bald eagle (FT), osprey, red-shouldered hawk, spotted sandpiper, ring-billed gull, Caspian tern, belted kingfisher, Eastern phoebe, northern rough-winged swallow, bank swallow, Louisiana waterthrush. MAMMALS: Beaver, mink, river otter.

DISTRIBUTION

Rivers occur throughout all of the continents, most prominently where precipitation is highest and the drainage basin broadest. The Mississippi is the only North American river among the world's 15 largest, ranking sixth. The largest river in New England is the Connecticut, a fifth- or sixth-order river (the Mississippi is a tenth- or twelfth-order river) 409 miles long (66 miles are in Massachusetts), and with a total drainage area of 11,250 square miles, 660 of them in Massachusetts. (The Mississippi is 2,348 miles long and drains 1,247,300 square miles.)

EXTENT IN MASSACHUSETTS

There are at least 2,027 named rivers and streams in Massachusetts; if laid end to end they would be at least 10,700 miles long. They belong to seven major river systems or drainage basins—the Hudson, the Housatonic, the Connecticut, the Thames, the Blackstone, the Merrimack, and the coastal systems combined. These can be further

Eastern phoebe fledglings

subdivided into twenty-nine individual river basins: the Hoosic, the Kinderhook, the Bashbish, the Housatonic, the Deerfield, the Westfield, the Farmington, the Connecticut, the Millers, the Chicopee, the Quinebaug, the French, the Nashua, the Blackstone, the Merrimack, the Shawsheen, the Assabet, the Sudbury, the Concord, the Parker, the Ipswich, the north coastal basin (comprising the Essex, Danvers, Pines, Annisquam, and Saugus rivers), Boston Harbor (comprising the Mystic, Neponset, and Weymouth rivers), the Charles, the south coastal basin (comprising the North, South,

and Jones rivers), the Taunton, the Ten Mile (Narragansett Bay basin), the Cape Code basin (Herring River), the Buzzard's Bay basin (comprising the Agawam, Wareham, Weweantic, Mattapoisett, Acushnet, Slocums, and Westport rivers).

CONSERVATION STATUS

Given the benefits we have reaped from our rivers, it is sad to report that river communities are the most degraded ecosystems in the United States. Since colonization began we have reduced riparian habitat in this country from 121 million acres within the hundred-year floodplain to 23 million, an 81 percent loss. The litany of crimes against our rivers is extensive. We have turned them into open sewers and conduits for industrial wastes, including toxic, environmentally persistent chemicals

Northern dusky salamander

Northern two-lined salamander

and heavy metals. We have clouded them with the eroded topsoil washed off the land that we cleared almost completely of soil-holding forest. We have dumped tons of deadly pesticides onto agricultural fields, much of which was washed into our rivers and ended up in the bodies of fish, bald eagles, and all other river organisms. At this writing, over 60 percent of Massachusetts rivers are unsafe for fishing and swimming (the national average is 30 percent), and river pollution threatens the public drinking water supplies of almost 4 million state residents who depend on reservoirs fed by rivers. We have dammed our rivers for drinking water, hydroelectric power, and recreation, thus altering many ecosystems and closing off natural pathways for anadromous fish and other riverine species—Atlantic sturgeon were once so abundant in the Merrimack that people were warned not to venture onto the river during their migrations, lest their boats be overturned by these gigantic fish. And we have grossly diminished our rivers' great beauty by building extensively along their banks.

The good news is that we have begun to recognize and atone for our sins. Strict federal and state environmental laws have curtailed many of the worst abuses, though of course much historical pollution still resides in bottom sediments. And rivers have a way of capturing people's imaginations, stimulating the formation of save-the-river campaigns: The success of the Nashua River Cleanup Committee in the 1960s stimulated other communities to organize and demand clean rivers from business and government. The Massachusetts Department of Fisheries, Wildlife, and Environmental Law Enforcement has instituted an "Adopt-a-Stream" program designed to assist people who want to improve and protect waterways in their communities. There are now over thirty watershed associations and other river-protection groups in the Commonwealth.

This kind of grass-roots conservation activism is heartening, but lest we be complacent it should be noted that as this book goes to press the Massachusetts legislature has failed to pass an effective river-protection bill, largely responding to political pressure from the real estate lobby.

PLACES TO VISIT

The amount of protected land along the banks of major Massachusetts rivers is notably small. Some exceptions are included below:

Maudsley State Park (Merrimack River), Newburyport. Massachusetts Department of Environmental Management (DEM).

Ipswich River Wildlife Sanctuary, Topsfield. Massachusetts Audubon Society (MAS).

Great Meadows National Wildlife Refuge (Concord River); Wayland, Sudbury, and Concord. U.S. Fish and Wildlife Service.

Broadmoor Wildlife Sanctuary (Charles River), South Natick. MAS.

Arcadia Wildlife Sanctuary (Mill River), Easthampton. MAS.

South Shore Sanctuary (North River), Marshfield. MAS.

Chesterfield Gorge (Westfield River), West Chesterfield. The Trustees of Reservations (TTOR).

Swift River Reservation, Petersham. TTOR.

Monroe State Forest (Dunbar Brook), Monroe. DEM.

FURTHER READING

Adopt-a-Stream Workbook, by the Riverways Program. Boston: Massachusetts Division of Fisheries and Wildlife, n.d.

Aquatic Entomology, by W. Patrick McCafferty. Boston: Science Books International, 1981.

An Atlas of Massachusetts River Systems, edited by Walter E. Bickford and Ute Janik Dymon. Amherst, Mass.: University of Massachusetts Press, 1990. An excellent overview of Massachusetts rivers with a wealth of information about protected areas and public access points.

The Connecticut, by Walter Hard. New York: Rinehart, 1947.

The Run, by John Hay. Garden City, N.J.: Doubleday & Co., 1959.

The Stream, by Robert Murphy. New York: Farrar Straus & Giroux, 1971.

Trout Streams, by Paul R. Needham. New York: Winchester Press, 1969.

Northern dusky salamander

Floodplain Forest

*The woods on the neighboring shore were alive with [passenger] pigeons,
which were moving south, looking for mast, but now, like ourselves, spending their noon in the shade.... We obtained one of these handsome birds,
which lingered too long upon its perch, and plucked and broiled it...for
our supper.*

HENRY DAVID THOREAU
A Week on the Concord and Merrimack Rivers

A floodplain is the area of flatland that is covered by water when a river reaches maximum height. The soils of the floodplain are composed largely of sediments that are picked up in the faster upper reaches of a river and then fall out of suspension as the river slows down over lower gradients. Quantities of organic material are also transported and deposited and, when mixed with varying textures of clays, silts, and sands, create dark, moist, fertile, usually deep soils without distinct horizons. The forest that grows out of these riverine soils typically consists of tall, straight, well-spaced trees that form a closed or partially open canopy. In Massachusetts the dominant tree species is usually silver maple, but cottonwood, sycamore, black willow, and other species are also characteristic. At the river's edge, where flooding is most frequent and prolonged, shrubs are sparse or absent, but a thick herbaceous layer develops once the spring flood has receded. Farther up the banks, on drier soil, a dense understory of shrubs and vines is typical.

The floodplain is naturally a very dynamic zone because of the constantly changing course, height, and velocity of the water. This is reflected in the forest structure by such characteristics as multiple trunks, caused by ice shearing of young trees; frequent tree falls, which then act as soil traps; changes in the composition of herb and shrub layers as soils are flooded or dry out; and the creation of topographical features such as pools, oxbow ponds, and sandbars, which support characteristic species of plants and animals. In some areas the river creates a series of terraces which become successively drier and support different plant communities as they climb the bank.

Because they follow water courses, floodplain forests tend to be long and narrow, though where the floodplain is broad they can extend well inland. These wooded corridors—which in New England tend to run north-south, following the glacier's path—may be important in the dispersal of forest wildlife such as migratory songbirds.

Perhaps the most enjoyable way to experience the floodplain forest is from a canoe. The pale shiny leaves and scaly bark of the silver maple

Opposite: The Concord River in late May

give this forest a distinctive shimmering aspect unlike that of any of our other forest types, and the giant sycamores or groves of statuesque cottonwoods that appear from time to time give a feel for the grandeur of the original river forests that once filled our valleys. At all seasons watch for bald eagles, red-shouldered hawks, and other raptors perching at mid-height where the trees face the river. Wood ducks and hooded mergansers, which arrive with the spring floods, are especially fond of the shady edges and forest pools. Warblers, thrushes, and other songbirds thrive in these insect-rich, many-layered forests. In summer listen especially for the sweet songs of warbling and yellow-throated vireos, which favor the canopies of riverside trees for nesting. And if you are lucky, a black-billed or yellow-billed cuckoo, species that also seem partial to this habitat, may cross the stream in front of you. Botanists will probably have to go ashore, where dense thickets of stinging nettle and poison ivy and legions of blood-thirsty mosquitoes can make a hot summer day in the intensely humid forest interior especially memorable. Rewards may include the exotic-looking green dragon and other rarities such as moonseed, winged monkey-flower, featherfoil, and narrow-leaved spring beauty.

INDICATOR SPECIES

Many species of plants and animals occurring in floodplain forest are also typical of wooded swamps and deciduous forest. Many of the rarest floodplain forest species occur only in the western part of the state. ❧ TREES: Silver maple, cottonwood, black willow, sandbar willow (SSC), sycamore, American elm, green ash, box elder, river birch (mainly Merrimack River), swamp white oak, pin oak (lower Connecticut River), bur oak (SSC; Housatonic River), butternut (higher terraces), American hackberry (higher terraces). ❧ SHRUBS AND VINES: Blackberries, wild black currant, riverbank grape, Virginia creeper, poison ivy, Morrow honeysuckle (X), Canada moonseed. ❧ GRASSES AND SEDGES: Riverbank rye, hairy wild rye (ST), Wiegand's wild-rye, tall brome, Frank's lovegrass (SSC; on open sandbars), Davis's sedge (SE; one of New England's rarest plants; Housatonic River), Gray's sedge (ST), hairy-fruited sedge (ST), Tuckerman's sedge (SE), cattail sedge (ST), foxtail sedge (ST), pubescent sedge, long-beaked sedge. ❧ WILDFLOWERS: Green dragon (ST), smooth Solomon's seal, wild garlic, stinging nettle (X), wood nettle, clearweed, false nettle, narrow-leaved spring beauty (ST), marsh yellowcress var. *fernaldiana*, rough avens, jewelweed, American germander, winged monkey-flower (SE), carpenter's square, wild cucumber, one-seeded bur cucumber, crooked-stemmed aster (SSC), great ragweed, cocklebur var. *canadense*, green-headed coneflower, sneezeweed. ❧ FERNS: Ostrich fern. ❧ BUTTERFLIES: Eastern comma (caterpillar feeds on elm, nettles, and hops). ❧ DRAGONFLIES: Beaked and fawn darners patrol shady areas along river banks and forest openings. For other species see Dragonflies in RIVERS AND STREAMS. ❧ AMPHIBIANS: Mole salamanders in vernal pools in forest interior. ❧ BIRDS: Red-shouldered hawk, wood duck, hooded merganser, black-billed and yellow-billed cuckoos, veery, warbling and yellow-throated vireos, blue-gray gnatcatcher. ❧ MAMMALS: Mink, raccoon.

KEY
SE=State Endangered
ST=State Threatened
SSC=State Special Concern
X=Exotic

Eastern comma

DISTRIBUTION

Floodplain forests grow worldwide wherever the appropriate conditions exist. In North America the great bottomland forests of the Mississippi and other large river systems are among the continent's major ecosystems. In Massachusetts, the best examples of this greatly diminished habitat can be seen along the Connecticut and its tributaries, as well as along the Housatonic, the Merrimack, and the Ipswich rivers.

EXTENT IN MASSACHUSETTS

Floodplain forest is not a recognized commercial forest type, and apparently no systematic survey has been attempted. It may be fair to estimate a total acreage for Massachusetts of less than (perhaps *considerably* less than) 10,000 acres.

CONSERVATION STATUS

Mature, undisturbed, diverse floodplain forests, once a major ecosystem of Massachusetts and southern New England, have been reduced by human activities to a few limited patches and strips. They have been converted mainly to agricultural fields, industrial uses, and housing developments, despite the obvious perils of living in a floodplain. Mills, hydroelectric dams, and flood-control measures have drowned many additional acres. In addition to the loss of biodiversity and aesthetic and recreational values, the destruction of these forests has resulted in increased erosion and flooding of human habitations, loss of water-storage capacity, degradation of drinking-water quality, and reduction of fisheries.

PLACES TO VISIT

Ipswich River Wildlife Sanctuary, Topsfield. Massachusetts Audubon Society (MAS).
Arcadia Wildlife Sanctuary (Connecticut River), Easthampton. MAS.
Canoe Meadows Wildlife Sanctuary (Housatonic River), Pittsfield. MAS.

FURTHER READING

An Atlas of Massachusetts River Systems, edited by Walter E. Bickford and Ute Janik Dymon. Amherst, Mass.: University of Massachusetts Press, 1990.
See also general forest references in OAK-CONIFER FOREST.

Cultural Grasslands

Eastern meadowlark

*F*ield, pasture, meadow, lawn, plain, prairie, steppe, pampa are all terms that describe grasslands. Some of them, such as *field*, are broadly descriptive; others are used more specifically. For example, some ecologists define a pasture as a grazed grassland and a meadow as a mown grassland, though such distinctions are by no means universal. And several of these names have a cultural origin. *Prairie* was the word used by French explorers of the seventeenth century to describe the tall grasslands they found in what is now the Midwestern United States. *Steppe* derives from a Russian word that originally described the great rolling plains that extend from Eastern Europe through Central Asia. And *pampas* is a Spanish coinage for the prairies of central Argentina. Whatever we call them and however they vary, all of these communities have one thing in common: They are dominated by one or more species of grass.

Natural grasslands—those not destined to succeed to forests if left alone—occur only where nutrient-poor soils, harsh climatic conditions, or large numbers of endemic wild grazing animals inhibit the growth of woody plants. A North American example is our western Great Plains, where only grasses and a few other hardy herbaceous plants can tolerate the relatively low rainfall, constant drying wind, and prolonged frosts that are characteristic of that region.

But in New England—where bare earth, left undisturbed, will eventually turn to forest—natural, self-sustaining grasslands are scarce and limited, except for coastal salt marshes, also known as salt meadows. Otherwise natural grasslands are

As I looked about me I felt that the grass was the country, as the water is the sea....And there was so much motion in it; the whole country seemed, somehow, to be running.

WILLA CATHER
My Ántonia

restricted to riverbanks and inundated wetlands (including beaver meadows), where periodic flooding combined with limited or poor soil prevents the growth of shrubs and trees, and to a few patches of coastal or floodplain sands on the islands south of Cape Cod and in the Connecticut River Valley (see SANDPLAIN GRASSLAND). The other fields, pastures, and meadows of Massachusetts were all created—most of them quite recently—by people, and must be maintained with the aid of livestock, machines, or fire. These are *cultural* grasslands.

It would be fascinating to know what kind of grasses and wildflowers dominated the clearings made by Native Americans or by natural wildfires in the Northeast before colonization. We do know that the settlers complained of the poor forage available and soon imported reliable Eurasian species both intentionally and inadvertently. These grasses and pasture weeds naturalized so quickly it is impossible to know exactly what a native eastern successional grassland was like. Today the average upland pasture with decent soil—even one that has been allowed to go weedy with wildflowers—is likely to consist of more than 50 percent nonnative grasses and forbs (herbaceous flowering plants other than grasses, sedges, or rushes). There are a few species that apparently were native to both the New World and the Old, but no botanists were present to record such fine points.

Grasses are enormously important to the human economy. About a third of the world's land area and half of the area of the United States is grassland. All of our food grains, including corn, wheat, rice, and oats, are grasses, as are sugarcane and bamboo. Without forage and feed grasses, the commercial production of cows, sheep, chickens, and other livestock would not be possible. And at least two forms of grassland, suburban lawns and golf courses, have become universal symbols of Western culture. Disregarding the ecological consequences of this vast anthropogenic transformation of the landscape—viewed variously as a splendid pioneering achievement or ruthless devastation at an unprecedented rate—it is clear that the forests that were cut, the game that was "harvested," and the agricultural grasslands that were created were essential to the survival and prosperity of the new Americans.

One exception to these Europeanized grasslands is what we call an *old field,* one that has been allowed literally to go to seed. Especially common in eastern Massachusetts, where the soils are often very thin, well-drained, and/or acid, these largely native communities are dominated by little bluestem, a grass species common in the Midwestern prairies, and are typically dotted with red cedar (juniper), perhaps because grazing animals avoided this plant's prickly, toxic foliage. On the richest sites old field gives way eventually to ericaceous (heath family) shrublands and oak woods. But where old fields persist, either naturally on poor sites or through human intervention, they contain fewer alien species and are essential habitat for several butterfly species, especially dusted and cobweb skippers. Where the junipers grow up densely they often support colonies of the exquisite little olive hairstreak.

Another good use for a time machine would be to go back and observe the movement of prairie species eastward as the forests fell and grassland

Red fox

corridors opened from the Midwest to New England. Many of our rarest grassland species (upland sandpiper, regal fritillary) as well as our most characteristic farmland birds (bobolink, vesper sparrow), and at least one aggressive pest (brown-headed cowbird) have their centers of abundance to the west and must have been very rare and local or absent in the East at the time of Columbus. Clearly they all reached their apogee in New England during the "prairie years" of the 1800s, when the landscape was more pasture than forest. Is their present decline, paralleling the return to the forest, simply a return to normal?

Characterizing the biota and ecology of our cultural grasslands is a little tricky, precisely because they are not natural communities. Some of their variations are due to natural factors such as soil types, but others relate more to what was planted when and what kinds of manipulations have been done since. Key criteria for the cultural grassland community as defined here are that it (1) occurs on upland sites (i.e., not wet meadows or grassy marshes); (2) is typically dominated by a mixture of Eurasian grasses and forbs, with some exceptions (old fields); (3) is maintained artificially, usually by mowing or grazing; and (4) frequently supports a highly diverse community of animal species: rodents and orthopteran insects feed on grasses; raptors and grassland songbirds require open habitats with dense vegetation for foraging

Woodchuck

and/or breeding; many butterflies and moths seek the abundant nectar sources; and other organisms (e.g., many birds, reptiles, and dragonflies) are attracted by the abundance of invertebrate food.

Whether they are "natural" or not, cultural grasslands now represent unique biological resources supporting many species that occur in no other habitat in Massachusetts. They have also become a habitat much favored by certain human subspecies. The story is told of a plainsman forced by circumstances to move to the mountains. His new home is situated in a valley of the Rockies with snow-covered peaks of unsurpassable splendor rising all around. A passing traveler who stops to ask directions expresses admiration for the astounding alps, now turning several colors in the light of the waning sun. "They're all right, I guess," replies the plainsman, tolerantly, "but they sure do block out the view."

Anyone familiar with the broad rolling pastureland crossed and bounded at intervals by swaths of woodland—the agricultural patchwork that can still be seen in parts of western Massachusetts and that many still think of as the typical New England landscape—will have some sympathy for the plainsman's complaint. There is something about a vista, the ability to see beyond a curtain of mountain or trees, that simultaneously quickens the senses and eases the soul.

And as if to affirm that our species occasionally has a positive impact on the environment, these manmade wilds are also magnets for wildlife. No one who has looked for butterflies in a June meadow or watched half a dozen short-eared owls coursing over winter pastures at dusk or jumped a fox in a field can imagine that we are alone in benefiting from these welcoming spaces.

INDICATOR SPECIES

Note the high percentage of exotic species in the plant list. ❧ TREES: Red cedar (characteristic of old fields), occasional old pines, oaks, elms, or other picturesque shade trees left for livestock. ❧ SHRUBS: Dwarf cedar (old field). Abandoned pastures are soon invaded by blueberries, viburnums, dogwoods, spireas, and other native shrubs and by aggressive exotic shrubs such as European buckthorn (X), multiflora rose (X), autumn olive (X), and Japanese barberry (X). ❧ GRASSES: Little bluestem (dominant in old fields), big bluestem (occasional in old fields). Most common pasture grasses include orchard grass (X), timothy (X),

KEY
SE=State Endangered
ST=State Threatened
SSC=State Special Concern
X=Exotic

154

English rye grass (X), sheep fescue (X), meadow fescue (X), sweet vernalgrass (X), Kentucky bluegrass, smooth brome (X), velvet grass (X), foxtail grass (X; several species), tall oatgrass (X), quackgrass, red fescue. Redtop and reed canary-grass are both native and exotic; they apparently occurred in relatively localized habitats before European settlement and did not become the pervasive field plants they are today until European forms of the same species were introduced as forage crops. In managed hayfields, timothy and other species may grow in a near monoculture; such fields are less diverse biologically than abandoned fields gone to splendid chaos. ✎ **WILDFLOWERS:** Only the most common *upland* field species are listed, including a great variety of naturalized weedy species—many of which are highly attractive to butterflies and moths as nectar sources—and many of our most familiar wildflowers. Field garlic (X), common blue-eyed grass and several similar species, slender and little ladies'-tresses, bastard toadflax, buckwheat (X), Japanese knotweed (X), lady's thumb (X), erect knotweed, sheep sorrel (X), curled dock (X), lamb's-quarters (X), amaranths or pigweeds (several exotic species), bouncing bet (X), Deptford pink (X), maiden pink (X), ragged robin (X), bladder campion (X), white campion (X), night-flowering catchfly (X), corn spurrey (X), corn-cockle (X), common chickweed (X), mouse-ear chickweed (X), lesser stitchwort (X), bulbous buttercup (X), hoary alyssum (X), field peppergrass (X), field pennycress (X), shepherd's purse, wild peppergrass, tower mustard, dame's rocket (X), wild radish (X), field mustard (or rape) (X), charlock, (X), Indian mustard (X), common winter cress (X), black mustard (X), hedge mustard (X), tumble mustard (X), 6 species of cinquefoils (native and exotic), common blackberry, dewberry (vine), wild strawberry, many clover species (mainly X), alfalfa (X), black medick (X), 5 or more species of exotic vetches, bird's-foot trefoil (X), partridge pea, several exotic species of cranesbills, storksbill (X), cypress spurge (X), leafy spurge (X), musk mallow, common St. John's-wort (X), ovate-leaved violet, bird's-foot violet (most open-country violets prefer wet meadows), common evening primrose, Queen Anne's lace (X), golden Alexanders, gentians, spreading dogbane, common milkweed, purple milkweed, butterfly-weed, black swallow-wort (X; vine), field bindweed (X), viper's bugloss (X), hound's tongue (X), European (X) and corn (X) gromwells, narrow-leaved vervain, several species of mountain mints, selfheal, wild thyme (X), American pennyroyal, hyssop (X), several species of henbits (X), eyebright, gill-over-the-ground, yellow rattle, horse mint, several native species of ground cherries, horse nettles, blue toadflax, butter-and-eggs (X), several species of native and exotic speedwells, slender gerardia, common mullein (X), wild madder (X), yellow bedstraw (X), common bluet, corn salad (X), Indian tobacco, hawkweeds and mouse-ears (several exotic species), common dandelion (X), fall dandelion (X), cat's ear (X), lamb succory (X), yellow goat's beard, several exotic species of sow thistles, smooth hawksbeard (X), ox-eye daisy (X), lance-leaved coreopsis (X), pussytoes (several native species), pearly everlasting, sweet everlasting, common burdock (X), common clotbur, tansy (X), black and brown and spotted knapweeds (all X), chicory (X), thistles (field, pasture, yellow, Canada [X], and bull [X]), Spanish nee-

Bobolink

dles, sticktight, hyssop-leaved boneset, common, lesser, and daisy fleabanes (all X), Robin's plantain, horseweed, common groundsel (X), pineapple weed (X), yarrow (X), common mugwort (X), common ragweed (X), field scabious (X), teasel (X), common plantain (X), English plantain (X), narrow-leaved white-topped aster, black-eyed Susan, goldenrods (especially rough-stemmed, Canada, gray, early, late, tall, lance-leaved, and slender-leaved), asters (especially New England, calico, many-flowered, heath, small white, bushy, and smooth). ∾ DRAGONFLIES: A number of species habitually feed in upland fields as adults: slender and other spreadwings; large numbers of common green darners roost in fields during fall migrations; mixed swarms of darners and emeralds patrol for insects over fields at dusk; common and prince baskettails hawk over grasslands at midday; eastern amberwing, calico pennant, Halloween pennant; most meadowflies (especially cherry-faced, ruby, band-winged, and yellow-legged). ∾ BUTTERFLIES: Tiger swallowtails, eastern black swallowtail, cabbage white (X), common sulphur, orange sulphur, American copper, great spangled fritillary, aphrodite fritillary, pearl crescent, Harris checkerspot, painted lady, American lady, red admiral, viceroy, inornate ringlet, large wood nymph, monarch, common sooty-wing. Most of our smaller skippers feed on grasses; the following species are the most strongly associated with upland fields: European skipper (X), Leonard's skipper, cobweb skipper (old fields), Indian skipper, yellow-patch skipper, tawny-edged skipper, crossline skipper, long dash, little glassy-wing, Delaware skipper, Hobomok skipper, dun skipper and dusted skipper (old fields). ∾

AMPHIBIANS: American toad, northern leopard frog, pickerel frog. ❧ **REPTILES:** Snapping turtle (sometimes buries eggs in upland fields), wood turtle (SSC), eastern box turtle (SSC), painted turtle (egg-laying), northern brown snake, northern redbelly snake, eastern garter snake, northern black racer, smooth green snake, eastern milk snake. ❧ **BIRDS:** Cattle egret (with livestock), Canada goose (grazes in mown fields), northern harrier (ST), red-tailed hawk (at field edges), rough-legged hawk (winter), American kestrel, ring-necked pheasant (X), black-bellied plover (migrant), American golden-plover (migrant), killdeer, upland sandpiper (SE), buff-breasted sandpiper (fall migrant), Baird's sandpiper (fall migrant), gulls (frequently roost in farm fields), short-eared owl (SE; other owls also forage over fields at least occasionally), common nighthawk, eastern kingbird, horned lark; all 6 of our swallows hawk for insects over fields; American crow, eastern bluebird (wooded edges), American robin, American pipit, European starling (X), indigo bunting (fields invaded by shrubs), field sparrow (fields invaded by shrubs), grasshopper sparrow (ST), vesper sparrow (ST), Lapland longspur (migrant), snow bunting (migrant), bobolink, eastern meadowlark, red-winged blackbird, common grackle, brown-headed cowbird. ❧ **MAMMALS:** Short-tailed shrew, hairy-tailed mole, eastern mole. Many bat species feed over fields and along woodland borders. Eastern cottontail, New England cottontail (prefers brushier habitat), wood-

chuck, meadow vole, meadow jumping mouse, eastern coyote, red fox (X), white-tailed deer.

DISTRIBUTION
Cultural grasslands are found worldwide except in the polar regions. They are now most widespread where farming and livestock grazing are still practiced. Southern New England was largely (ca. 75 percent) deforested and converted to farmland by

the mid-nineteenth century. This trend and percentage is now reversed, agriculture continues to decline in the region, and with it the area of cultural grassland.

EXTENT IN MASSACHUSETTS
Three types of landscape that might be described as cultural grasslands of significant value to wildlife were defined and quantified by MacConnell (1975). The first, "Unused tillable land...not recently tilled and not part of an agricultural unit...occurring near growing urban areas and...usually mowed annually to maintain its value" amounted to approximately 13,820 acres as of 1971–72. The second type, "Pasture or wild hay land...not suitable for tillage due to steepness of slope, poor drainage, stoniness, or lack of fertility...often has scattered shade trees for the grazing

animals" was calculated to total approximately 168,671 acres. Finally, "Abandoned field…reverting to wild land, woody vegetation and grass abundant, but tree crown cover less than 30%…" largely "pasture or wild hay land before abandonment" defines a third type that accounted for approximately 164,630 acres. The total acreage in these categories in 1971–72 was 347,121 acres, or about 6.7 percent of the total area of Massachusetts.

Active agricultural land, that is, "tilled or tillable cropland which is or has recently been intensively farmed (including farm buildings)" constituted an additional 241,673 acres as of 1971–72.

Airports, which are typically maintained in a combination of short-mowed fields and expanses of tall grass and weeds, and which are little disturbed by people on foot, have become significant islands of habitat for some of our rarest grassland birds such as the upland sandpiper. Total airport land in 1971–72 (including buildings) was 10,805 acres.

CONSERVATION STATUS

Cultural grasslands present a number of interesting problems for conservationists. They are by definition artificial habitats, yet they sustain a community of species, including a number of rarities, that cannot survive—or can do so only marginally—outside of them. They were originally created and maintained by agricultural practices, whose goal was to grow food and forage, not to create habitat. Many of these practices—the creation of large monocultures, heavy applications of fertilizers and pesticides, the harvesting of hayfields before nestling birds have fledged, pollution

of wetlands and waterways from runoff and animal manure—have a detrimental impact on the environment. But maintaining "natural" grasslands for reasons of biological diversity is expensive and time-consuming and yields little (e.g., late-cut hay) to offset the expense.

There is no question that in New England large tracts of grassland, whether active farmland or other open space maintained in a way that promotes biological diversity and the presence of rare species, are now scarce and dwindling. In many cases farmers who can no longer afford to farm are forced to sell their land—often well-drained uplands that make ideal residential or industrial sites—to developers. The approach to maintaining grasslands that seems to be evolving in Massachusetts at present has three basic components: (1) The Massachusetts Natural Heritage and Endangered Species Program, the Nature Conservancy, and Massachusetts Audubon's Grassland Conservation Program have identified the most important grassland sites in the state and continue to catalogue other grasslands with significant assemblages of species. These become prime candidates for protection by conservation organizations or government agencies. (2) Ecologists are becoming increasingly sophisticated in managing grasslands in ways that maximize their ecological value at relatively low cost. The dissemination of these methodologies to landowners who wish to preserve grasslands for ecological and aesthetic reasons creates additional acres of this habitat. (3) Constituencies such as conservationists, farmers, and the military, who at first glance would appear to have little in common, are finding that there are mutual benefits to managing grasslands in ways

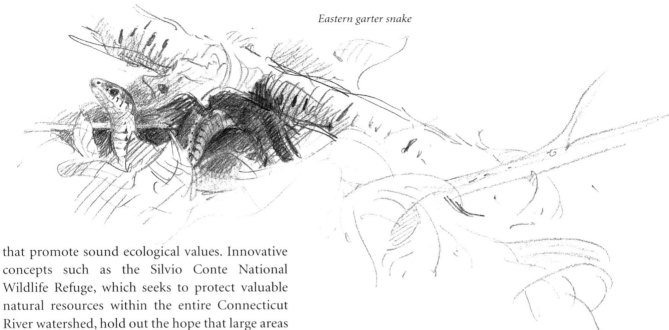

Eastern garter snake

that promote sound ecological values. Innovative concepts such as the Silvio Conte National Wildlife Refuge, which seeks to protect valuable natural resources within the entire Connecticut River watershed, hold out the hope that large areas of habitat can be preserved and managed through creative partnerships rather than by the traditional methods of land acquisition.

PLACES TO VISIT

Daniel Webster Wildlife Sanctuary, Marshfield. Massachusetts Audubon Society.

The Common Pastures, West Newbury. Mainly privately owned, accessible by road.

World's End Reservation, Hingham. The Trustees of Reservations (TTOR).

Bartholomew's Cobble, Ashley Falls. TTOR.

Moran Wildlife Management Area, Windsor. Massachusetts Division of Fisheries and Wildlife.

Cumberland Farm Fields, Middleboro. Access from Rivers Street; privately owned.

See also listings in SANDPLAIN GRASSLAND.

FURTHER READING

Grasses: An Identification Guide, by Lauren Brown. Boston: Houghton Mifflin, 1979.

Grasslands, by Lauren Brown. New York: Knopf, 1985. An Audubon Nature Guide.

The Life of Prairies and Plains, by Durward L. Allen. New York: McGraw-Hill, 1967.

Prairie Plants and Their Environment: A Fifty Year Study in the Midwest, by J. E. Weaver. Lincoln, Neb.: University of Nebraska Press, 1968.

Remote Sensing: 20 Years of Change in Massachusetts 1951/52–1971/72, by William P. MacConnell. Research Bulletin #630, Amherst, Mass.: Massachusetts Agricultural Experiment Station, 1975.

Old Growth Forest

This is the forest primeval. The murmuring pines and the hemlocks,
Bearded with moss, and in garments green, indistinct in the twilight,
Stand like druids of old with voices sad and prophetic....

HENRY WADSWORTH LONGFELLOW
Evangeline

Old growth forest is not a type of natural community but rather a condition or quality that can prevail in any forest community. But examples are so rare in Massachusetts—or indeed anywhere in the Northeast—that the phenomenon merits a profile of its own.

Aficionados of old growth—enough of whom now exist to hold conferences on the subject—have many ideas of what, exactly, constitutes old growth forest. However, all agree that the term best applies to forests in which trees have matured and died over several generations with minimal human disturbance. This definition has been further refined in different areas. For example, in his survey of old growth forest in Massachusetts, Peter Dunwiddie (1992, 1995) identified four criteria for defining the phenomenon. To qualify, a stand must be

- Free of any sign of human or other catastrophic disturbance. This excludes sites with features such as roads, stone walls, or abundant charcoal (indicating a stand resulting from a fire), but allows for the occasional trail.
- Of sufficient area (at least ten acres in the Northeast) to constitute a forest, rather than just a few old trees.
- Relatively stable in composition and structure, including late-successional, or climax, species that are successfully reproducing within the stand, creating a variable-aged forest.

- Dominated by individual trees of some minimum age. This varies between species and sites, but attaining half the maximum age of any given species seems to be a credible criterion for venerability among forest Methuselahs. For sugar maple, beech, and hemlock this means a minimum of 150 to 200 years, with older individuals occasionally reaching 250 to 300 years or more.

Surprisingly, most of the old growth remnants that still exist in Massachusetts have gone unrecognized until recently. This may be due in part to the failure of these stands to fit our mental picture of what an ancient forest is *supposed* to look like: massive trunks rising like cathedral buttresses out of a moss-covered boneyard of fallen giants; shafts of light filtering down through a canopy 200 feet overhead; carpets of rare wildflowers and a better-than-average chance of glimpsing an archetypal forest denizen such as a bear. Some of our oldest oak-conifer and northern hardwood stands approach this kind of majesty with trunks of the largest trees measuring 10 feet or more around, rising 60 feet before the first branch and topping out at over 150 feet. And there is no shortage of light shafts and moss in these places. But more often, Massachusetts old growth stands are decidedly unimpressive, comprised of a few gnarled

Opposite: Old growth eastern hemlock

specimens of unspectacular height and girth amidst a host of younger trees. Such forests, often growing in poor soils on steep, rocky ridges and slopes exposed to the wind, have survived in part because of these adversities. Trees grow slowly under such conditions, and even after centuries, often fail to reach the massive size of those rooted in rich soils and in low areas more sheltered from storms. Trees in such sites that do manage to attain imposing stature are likely to be toppled by the high winds that blow across New England from time to time. Furthermore, the oldest trees frequently exhibit twisted trunks, rotten centers, broken limbs, and short boles, making them undesirable for timber and rendering them unlikely to be cut. These characteristics, combined with the inaccessibility of the precipitous terrain and the unsuitability of these sites for other uses, help preserve the forests growing there.

If most Massachusetts old growth stands are far from Bunyanesque, neither have they proven to be particularly rich in species or rarities. Indeed, most of the unique aspects of Massachusetts old growth seem to be structural. After a certain age the bark characteristics of trees change dramati-

cally; forest litter is notably deep; and the accumulation of a moist, organic-rich substrate over decades may allow some species—such as mosses that grow best on decaying wood—to thrive to a degree not possible in younger woodlands.

It is only fair to add that we still know relatively little about these vestigial forests, which have been brought to public and scientific attention only within the last decade. It may be that species of beetles, lichens, or mushrooms endemic to certain types of old growth forest have yet to be identified.

The fact that most of our old growth sites are not cathedral groves or teeming with biodiversity should not make us value them less. Most have a unique aesthetic appeal that is hard to define in biological terms, and for most people there is an inherent *spiritual* grandeur that adheres to these woods simply because they are so very old.

INDICATOR SPECIES
Were there sow bugs, bark beetles, and tardigrades endemic to New England's primeval forests that disappeared with the clearing of the land? Given that we have extirpated several species of forest birds since European settlement, it is certainly possible, but we will never know. No obligate inhabitants of New England old growth have yet been described, but investigation of the habitat is in its infancy. For characteristic species of most of our old growth forests, see OAK-CONIFER FOREST and NORTHERN HARDWOOD FOREST.

DISTRIBUTION
Old growth forest occurs worldwide in the few remaining places where forests have escaped cutting and other pervasive forms of anthropogenic dis-

Eastern chipmunk

turbance. "Virgin" stands are very rare anywhere in New England. In Massachusetts, old growth as defined above is now limited as far as we know to the Berkshires. The most significant sites include: a stand near the confluence of the Deerfield and Cold rivers; the Hopper on Mount Greylock (near the Deer Hill Trail); fragments in the watersheds of Fife, Dunbar, and Bash Bish brooks; parts of the Mohawk Trail State Forest; and the east side of Mount Everett.

EXTENT IN MASSACHUSETTS

About 300 acres of Massachusetts forest in about a dozen sites meet all of the criteria cited above. The largest sites are 50 to 75 acres. Most of these stands are dominated by hemlock or include this long-lived species as a major component. Perhaps another 2,500 acres may be described as "near old growth," that is, lacking one or more of the defining characteristics.

CONSERVATION STATUS

The human propensity to regard forests as a resource to be harvested explains the present rarity of old growth forest anywhere in the developed world. Even today traditional foresters are likely to view old growth as just so much overripe timber. And, despite an abundance of less quantifiable virtues, the fact that our ancient forests are not brimming with endangered species or charismatic megafauna makes them hard to sell as jewels in the crown of our natural heritage. Fortunately, several of Massachusetts's best sites are on protected land, and an effort is under way to include one or more of them in the state Nature Preserves system, giving them additional protection.

In addition, the Commonwealth has a great deal of middle-aged forest, which, though it can never be primeval, can eventually become ancient. From a biological conservation perspective, it seems obvious that Massachusetts should maintain a few extensive stands of undisturbed old growth forest, ideally including representatives of different forest types. But this notion has not been embraced by many conservation land owners. In its effort to maintain the broadest possible natural biological diversity, the Massachusetts Audubon Society has identified large tracts of mature forest in its sanctuary system for monitoring and management as eventual old growth stands.

PLACES TO VISIT

Mount Greylock Reservation, Berkshire County; access from North Adams and Lanesboro. Massachusetts Department of Environmental Management (DEM).

Mohawk Trail State Forest, Charlemont. DEM.

FURTHER READING

"Old Growth Forests of Southern New England, New York, and Pennsylvania," by Peter Dunwiddie, David Foster, D. Leopold, and Robert Leverett. In *Eastern Old Growth Forests: Prospects for Rediscovery and Recovery,* edited by M. B. Davis. Washington, D.C.: Island Press, 1995.

"Survey of old growth forests in Massachusetts" by Peter Dunwiddie. Final Report to the Massachusetts Natural Heritage and Endangered Species Program. Boston: Commonwealth of Massachusetts, 1992.

Northern Hardwood Forest

October foliage in Florida, Massachusetts

There is in some parts of New England a kind of tree...whose juice that weeps out of its incisions, if it be permitted slowly to exhale away the superfluous moisture, doth congeal into a sweet and saccharin substance, and the like was confirmed to me by the agent of the great and populous colony of Massachusetts.

ROBERT BOYLE, 1663

The northern hardwood forest is the northernmost of our deciduous forest communities. Its distribution closely matches a climate zone slightly warmer than that in which the spruce-fir forest typically occurs, one in which there are about fifty more frost-free days in a given year. It thrives best on soils that are well drained but remain moist year-round with a substantial 12- to 35-inch layer of fine glacial till, an unstratified mix of clay, sand, and gravel. The mineral content of the soil is also an important factor, and there is evidence that northern hardwoods grow best on a bedrock of schist. As a botanical term, *hardwood* refers to broad-leaved, flower-bearing (angiosperm), often deciduous trees, as distinguished from *softwoods,* which are usually needle-leaved and evergreen coniferous (gymnosperm) trees. *Hardwood* and *softwood* are essentially lumber terms that refer to the density of the wood. There are relatively soft hardwoods, such as poplar, and relatively hard softwoods, such as hemlock.

Dominant tree species of northern hardwood forests are sugar maple, American beech, yellow birch, and eastern hemlock, but a number of other hardwoods also occur commonly, possibly evidence of some form of natural or human disturbance. Except for yellow birch, all of these dominants are slow-growing trees that will sprout and grow in the shade of other species and eventually outcompete them as the forest matures. Another way of identifying this forest type is to observe which tree species are *not* present. Absent will be oaks (except for the occasional giant red oak), hickories, American chestnut, and tulip tree, typical of warmer, drier forests, as well as spruces and balsam fir, which start to intrude as conditions become colder and wetter.

Classic, mature northern hardwood forest has fewer species of plants and animals than either rich mesic forest or many mixed successional forests.

The northern hardwood forests constitute one of the world's most handsome forest types: stately, with mature trees topping 100 feet, and relatively open, with shafts of sunlight penetrating the canopy. This forest's surpassing glory is the color display it puts on each year in October, when the dying foliage bursts phoenixlike into flames of yellow, orange, and crimson, set off by dark blue-green patches of hemlock. A second worldwide claim to fame is of course the various sweets produced by one of the dominant tree species.

INDICATOR SPECIES

❧ TREES: Sugar maple (D), American beech (D), yellow birch (D), eastern hemlock (D), white ash, black birch, white birch, red oak, black cherry, red maple, American basswood, and white pine. Typical understory trees are striped maple, mountain maple, hornbeam, and witch hazel. ❧ SHRUBS: American yew, hobblebush, maple-leaf viburnum, red elderberry, partridgeberry (ground cover), American fly honeysuckle. ❧ WILDFLOWERS: Carolina spring beauty, troutlily, round-leaved violet, whorled aster, white wood aster, trilliums, especially red trillium. ❧ FERNS: Evergreen wood fern, Christmas fern. Several clubmosses are also typical. ❧ BUTTERFLIES: Canadian tiger swallowtail (caterpillar feeds on aspen and birch), West Virginia white (toothworts), veined (or mustard) white (SSC; toothworts), early hairstreak (ST; near obligate; American beech and beaked hazelnut), green comma (birch), gray comma (currants), Compton's tortoiseshell (birch), white admiral (birch). ❧ AMPHIBIANS AND REPTILES: Mole salamanders and other forest-inhabiting species of frogs, snakes, and terrestrial turtles may be found in northern hardwood forest, but none are strongly characteristic; red efts are notably common in this forest type. ❧ BIRDS: No obligates.

KEY
D=Dominant
ST=State Threatened
SSC=State Special Concern

Solitary vireo

166

Northern goshawk, yellow-bellied sapsucker, solitary vireo, black-throated blue warbler, and Louisiana waterthrush tend to prefer mature deciduous northern or mountain forest. For other widespread deciduous and mixed-forest species that may occur in northern hardwood forests, see OAK-CONIFER FOREST. ❧ MAMMALS: Smoky shrew, northern flying squirrel, woodland jumping mouse, porcupine, black bear, fisher.

DISTRIBUTION

Hardwood forests containing species of beech, birch, and maple occur in appropriate northern climates and corresponding altitudes in the mountains below the boreal spruce-fir forest community throughout the north temperate zone. In North America hardwood forest dominated by sugar maple extends as far west as Manitoba. However, the sugar maple–beech–yellow birch community described here as northern hardwood forest ranges from eastern Wisconsin and upper Michigan through southern Ontario and Quebec to extensive areas in the Maritime Provinces, south into Ohio and Indiana, and in isolated pockets in the Appalachians as far as West Virginia. In New England it extends mainly from south-central Maine through northern New Hampshire and most of Vermont to the highlands of western Massachusetts. In the Commonwealth most of our northern hardwood forest is in northern Berkshire and western Franklin counties. In the northern part of its range this forest type covers mountain slopes from near sea level to about 2,500 feet, whereas in the southern Appalachians it *begins* at 2,500 feet and reaches 4,500 feet.

EXTENT IN MASSACHUSETTS

According to Dickson and McAfee (1988), Massachusetts contained a total of 258,700 acres of the purest sugar maple–beech–yellow birch association as of 1985. Defined somewhat more broadly to include significant numbers of nondominant tree species, the total for all northern hardwood forest in the Commonwealth in 1985 was 671,000 acres.

CONSERVATION STATUS

As with all other New England forests, the most significant cultural influence on the northern hardwoods was the nearly total and often repeated clearing by European colonists. Soils, seed sources, the range of potentially competitive species, and disturbance regimes have all been altered by human influence, and introduced pests and diseases are vectors of continuing change. For example, the beech-bark fungus carried by an Asiatic aphid is taking its toll on New England beeches, thus altering the structure and composition of northern hardwood forests. Some studies suggest that atmospheric influences such as acid deposition, caused by the burning of fossil fuels, may be at least partly responsible for the degeneration of certain populations of sugar maples.

On the positive side, much of the regenerated northern hardwood forest in Massachusetts is contained within state parks and private conservation holdings.

PLACES TO VISIT

Mount Greylock State Reservation, Berkshire County; entrances in North Adams and Lanesboro. Massachusetts Department of Environmental Management (DEM).

Notchview Reservation, Windsor. The Trustees of Reservations.

Monroe State Forest (especially Dunbar Valley Backcountry Area), Monroe. DEM.

Pleasant Valley Wildlife Sanctuary, Lenox. Massachusetts Audubon Society.

Small areas of northern hardwoods occur at or near the summits of the taller hills of central Massachusetts, such as Mount Wachusett.

FURTHER READING

Deciduous Forests of Eastern North America, by E. Lucy Braun. New York: Macmillan, 1950.

A Field Guide to Eastern Forests, by John Kricher and Gordon Morrison. Boston: Houghton Mifflin, 1988.

Forest Cover Types of the United States and Canada, edited by F. Y. Eyre. Washington, D.C.: Society of American Foresters, 1980.

Forest Statistics for Massachusetts—1972 and 1985, by David R. Dickson and Carol L. McAfee. Broomall, Pa.: Northeastern Forest Experiment Station, USDA Forest Service, 1988.

A Natural History of Trees of Eastern and Central North America, by Donald Culross Peattie. Boston: Houghton Mifflin, 1948; reprint 1991.

Downy woodpecker

Rich Mesic Forest

The formative characteristics of the rich mesic forest community are sweet, or basic, soils enriched by calcium and moisture. (The word *mesic* means "moist," and mesic soils are neither very wet—hydric—nor very dry—xeric.) This combination allows increased microbial activity, which in turn leads to rapid decomposition of leaf litter and rapid nutrient cycling. Such rich soils are able to support an unusually diverse plant community, and the calcium content also favors particular species. Sugar maple is always the dominant tree species in this community, sometimes to the exclusion of all others. It has been shown that sugar maple leaves themselves are highly basic and therefore break down rapidly in soil formation, which in other types of forest is retarded by high acidity. This results in a thinner layer of leaf litter on the forest floor than in other forest types, which allows the earliest spring ephemeral wildflowers to emerge and get the sunlight they need before the trees leaf out.

Rich mesic forest tends to occur in places where there is calcareous bedrock near the surface or on slopes below calcareous outcrops.

May is the month to visit these rich woodlands. Before the maples and birches have unfurled their leaves and while the threat of a late snowstorm still remains, the forest floor erupts with a profusion of so-called early spring ephemerals—hepaticas, bloodroot, Dutchman's breeches, lilies, and others—that will disappear completely after their brief exposure to the sun.

Jack-in-the-pulpit

Long-spurred violet

Miterwort

INDICATOR SPECIES

Herbaceous species are listed in approximate order of decreasing association with the community, following Weatherbee and Crow's approach (see FURTHER READING). ❧ TREES: Sugar maple (always dominant), white ash, American basswood, bitternut hickory, black birch, yellow birch, American beech. Hop hornbeam is abundant as an understory tree. ❧ SHRUBS: Leatherwood. ❧ WILDFLOWERS: Blue cohosh, sharp-lobed hepatica, wild leek (ramps), wild ginger, common and cutleaf toothworts, squirrel corn, Dutchman's breeches, large-flowered bellwort, Virginia waterleaf, long-spurred violet, Selkirk's violet, showy lady's-slipper (SSC), yellow lady's-slipper (var. *pubescens*), hairy wood mint (SE), ginseng (SSC), long-fruited snakeroot, clustered snakeroot (ST), barren strawberry (SSC), early meadow rue, rue anemone, small-flowered crowfoot, broad waterleaf (SE), Canada violet, zigzag goldenrod, bloodroot, white baneberry, red baneberry, wood nettle, sweet cicely. ❧ FERNS: Maidenhair fern, maidenhair spleenwort, Goldie's fern, narrow-leaved spleenwort, rattlesnake fern. ❧ GRASSES AND SEDGES: Plantainlike sedge, Hitchcock's sedge (SSC), woodland millet (ST), rice-grass, bottlebrush grass, Dewey's sedge, ambiguous sedge, roselike sedge, hairy-leaved sedge. ❧ BUTTERFLIES: West Virginia white (caterpillar feeds exclusively on toothworts), veined (or mustard) white (SSC; caterpillar feeds on toothworts among other mustards). ❧ VERTEBRATES: No vertebrates are strongly associated with this forest type to the exclusion of others. For the many forest species that may occur, see vertebrate categories in OAK-CONIFER FOREST and NORTHERN HARDWOOD FOREST.

DISTRIBUTION

Rich mesic forest of the type described here occurs in patches with the necessary calcareous geology from the Great Lakes states east to New England. In New England it generally occurs within the distribution of northern hardwood forest. The best examples in Massachusetts are in Berkshire County, but other sites exist in Franklin, Hampshire, and Hampden counties.

EXTENT IN MASSACHUSETTS

Not recorded.

CONSERVATION STATUS

Globally, this community appears to be relatively secure, though it may be locally rare. Fortunately, many of the best Massachusetts rich mesic forest sites occur on protected land. These areas are botanical treasure troves, and while unscrupulous collecting of plants from the wild is not a major conservation issue in New England, the threat cannot be completely ignored.

KEY
SE=State Endangered
ST=State Threatened
SSC=State Special Concern

PLACES TO VISIT

Mount Greylock State Reservation, Berkshire County; entrances in North Adams and Lanesboro. Massachusetts Department of Environmental Management.

High Ledges Wildlife Sanctuary, Shelburne Falls. Massachusetts Audubon Society (MAS).

Field Farm, Williamstown. The Trustees of Reservations (TTOR).

Pleasant Valley Wildlife Sanctuary, Lenox. MAS.

Chapelbrook Reservation, South Ashfield. TTOR.

FURTHER READING

"The Natural Plant Communities of Berkshire County, Massachusetts," by Pamela B. Weatherbee and Garrett E. Crow. *Rhodora,* 94 (1992), no. 878, pp. 178–180.

Large-flowered trillium

Bog

After making my way through some pine groves and alder scrub I came to the bog. No sooner had my ear caught the hum of diptera around me, the gutteral cry of a snipe overhead, the gulping sound of the morass under my foot, than I knew I would find here quite special arctic butterflies.…And the next moment I was among them. Over the small shrubs of bog bilberry with fruit of a dim, dreamy blue, over the brown eye of stagnant water, over the flower spikes of the fragrant bog orchid (the nochnaya fialka *of Russian poets), a dusky little fritillary bearing the name of a Norse goddess passed in low, skimming flight.*

VLADIMIR NABOKOV
Speak, Memory

The classic bog is a roughly circular body of standing water, lined with a deep layer of ever-accumulating peat and rimmed with a floating, spongy, often "quaking" mat of sphagnum (peat) moss, which in turn supports a plant community dominated by low shrubs in the heath family and a few highly specialized, acid-tolerant herbs. The outer margins may be lined with taller shrubs and small trees. A bog may begin as an impermeable basin, such as a glacial kettlehole or clay-lined depression filled with rainwater or groundwater and having neither inflow nor outflow. The lack of flow results in low oxygen levels, minimizes the addition of mineral nutrients and prevents flushing. In this suffocating, highly acid (by definition, a bog has a pH of less than 4.2—typically around 3), nutrient-poor water, few bacteria or animals can survive. Consequently, the natural decomposition process is very slow, and dead sphagnum and other plant material accumulate faster than they decay, creating a deepening bed of peat in which bog plants can eventually take root. Because of their natural preservative qualities, bogs were widely used in northern Europe as burial sites by Iron Age people and have yielded up many corpses of great anthropological interest in excellent condition.

Bogs can be slow successional communities, with the peat eventually able to support a characteristic coniferous forest of black spruce and American larch (see SPRUCE-FIR FOREST). However, such succession may end in a variety of different plant associations, or may be reversed.

To the bryologically unenlightened, the species of the genus *Sphagnum* will tend to blur into a single soggy carpet. But in fact there are about twenty species of these handsome mosses in New England, and an experienced moss watcher can accurately guess the degree of wetness, acidity, and sun exposure in a bog by noting where the fussier species occur.

Despite the characteristically depauperate plant

Opposite: Pitcher plant

Chalk-fronted skimmer

and animal communities, a good bog is a naturalist's dreamland: often secluded by a surrounding wall of forest; the vegetation, water, and reflections laid out in a pleasing arrangement of planes and horizons; an unparalleled assortment of greens spotted with orchid pink in spring and splotched with moorland crimson in the fall; an assemblage of strange plants such as sundews and pitcher plants that must catch and digest insect food to survive; great multicolored dragonflies patrolling the water's edge; perhaps the ethereal piping of a hermit thrush floating on the spice-scented air; and of course the magic of walking on waves as you cross the undulating mossy mat.

The typical Massachusetts bog is technically known as a level bog. There are many other variations on the bog theme, such as raised bogs and sloped bogs. See also CALCAREOUS FEN and AT-LANTIC WHITE CEDAR SWAMP.

INDICATOR SPECIES
Few species thrive in the harsh chemical conditions of a bog; conversely, many bog species are "obligates," unable to survive in milder wetlands. ∾ TREES: Black spruce, tamarack. ∾ SHRUBS: Leatherleaf and swamp loosestrife colonize the edge of the water and form the structure to which the sphagnum mosses can attach and spread. Bog laurel, bog rosemary, Labrador tea, rhodora, large cranberry, small cranberry, dwarf huckleberry, mountain holly, sweetgale, black crowberry. ∾ MOSSES: Sphagnum species. ∾ GRASSES AND SEDGES: White beaksedge, silvery bog-sedge, slender woolly-fruited sedge, mud sedge, pendant bog-sedge, 3 species of cottongrass, twigsedge. ∾ WILDFLOWERS: Podgrass (ST), wild calla, three-leaved Solomon's seal, round-leaved sundew, spatulate-leaved sundew, horned bladderwort, pitcher plant, rose pogonia, grass pink, white-fringed orchis, yellow-eyed grass. In bogs with more nutrients the plant community may be enriched by additional species typical of closely related fens. ∾ DRAGONFLIES: Bogs are notably rich in odonate species, especially the darners, cordulines, and whitefaces: eastern red damsel, southern bog bluet, boreal bluet, northern bog bluet, sphagnum sprite, dusky clubtail, harlequin darner, Canada darner, obscure darner, Amazon darner. Banded bog skimmer (SE) and black bog skimmer (SE) are both globally rare New England endemics. Petite emerald, club-tailed emerald, greater ringed emerald, lesser ringed emerald, delicate emerald, broom-tailed emerald, common emerald, beaver-pond bottletail, elfin skimmer, pale-backed skimmer, four-spotted skimmer, black meadowfly,

KEY
SE=State Endangered
ST=State Threatened
SSC=State Special Concern

frosted whiteface, crimson whiteface, little spotted whiteface, allied whiteface. ∾ **BUTTERFLIES:** Bog copper (caterpillar feeds on leaves of large cranberry). ∾ **OTHER INVERTEBRATES:** Sphagnum cricket, several species of backswimmers and water boatmen (true bugs), whirligig and ground beetles. A number of moths are dependent on the pitcher plant, and at least 20 other moth species, including the bog tiger moth, the bog holomelina, the sundew cutworm, and the cranberry blossomworm feed on typical bog plants. At least one mite and one aphid are also bog-dependent. ∾ **FISHES:** Typically absent, but northern redbelly dace (SE) prefers acid habitats and smallmouth bass can also occur. ∾ **AMPHIBIANS:** Four-toed salamander (SSC) is closely associated with sphagnum, laying its eggs in cavities under clumps of this moss. Several frog species, especially spring peeper and wood frog, are acid-tolerant and sometimes occur

Chalk-fronted skimmer

in bogs, but are not reliable indicators. ∾ **REPTILES:** No indicators; common species may occur casually. Bog turtle (SE) prefers fens. ∾ **BIRDS:** Many spruce-fir forest and other boreal birds feed in bog openings. Among those that may be said to be typical of bogs are common snipe, olive-sided flycatcher, yellow-bellied flycatcher, veery, Lincoln's sparrow, rusty blackbird, and many other boreal species whose nesting range is mainly north of Massachusetts. ∾ **MAMMALS:** Smoky shrew, northern bog lemming, southern bog lemming (SSC).

DISTRIBUTION

Bogs occur worldwide predominantly in temperate—especially boreal—regions at high latitudes and altitudes. Canada, Russia, and Alaska are particularly rich in this ecosystem. In Massachusetts the classic bog described here is most frequent in the northern and western parts of the state, especially at higher altitudes, but similar peatlands occur statewide in many situations, including coastal dunes.

EXTENT IN MASSACHUSETTS

The Massachusetts Natural Heritage and Endangered Species Program estimates that there are only about forty bogs of the classic kettlehole type described here in the Commonwealth. Even if the definition is expanded to include bog communities that occur in bays of lakes, headwater swamps, and wooded swamps, the total area is less than 1,000 acres.

CONSERVATION STATUS

Globally, first-class (biologically diverse) bogs are regarded as vulnerable because their range is restricted and there are a number of threats to their biological integrity. In many areas bog peat is dug and sold commercially as fuel, though this has never been a major threat in Massachusetts, owing to the relative scarcity of the resource and the illegality of mining wetlands in this state. Many bog areas on the southeastern coastal plain have been converted to commercial cranberry production, and these operations frequently apply insecticides to protect their crops. There is also a thriving industry in the harvest and sale of sphagnum (peat moss) for horticulture, though it is illegal to harvest it in Massachusetts; most peat moss used in New England comes from Canada and Europe. The rarest and oddest bog plants—especially orchids, pitcher plants, and sundews—are sometimes pillaged by unscrupulous collectors.

Bogs and their inhabitants are fairly well protected in Massachusetts. An exception is former commercial cranberry bogs that have reacquired their natural diversity, and then are proposed for "restoration."

PLACES TO VISIT

Black Pond Bog, Cohasset. The Nature Conservancy.

Ponkapoag Bog, Blue Hills Reservation, Milton. Metropolitan District Commission.

Otis State Forest, Otis; by interpretive program only. Massachusetts Department of Environmental Management.

Ward Reservation Bog, North Andover. The Trustees of Reservations.

Boreal bluet

FURTHER READING

Bogs of the Northeast, by Charles W. Johnson. Hanover, N.H., and London: University Press of New England, 1985.

The Bog People: Iron Age Man Preserved, by P. V. Glob. Ithaca, N.Y.: Cornell University Press, 1969.

Carnivorous Plants, by A. Slack, Cambridge, Mass.: MIT Press, 1980.

Ecology of the Northern Lowland Bogs and Conifer Forests, by J. A. Larsen. New York: Academic Press, 1982.

The Ecology of Peat Bogs of the Glaciated Northeastern United States: A Community Profile, by Antoni W. H. Damman and Thomas French. U.S. Fish and Wildlife Service, Biological Report 85. Washington, D.C.: GPO, 1987.

Ecosystems of the World. Mires: Swamp, Bog, Fen and Moor, edited by A. J. P. Gore. Amsterdam: Elsevier, 1983.

Field Guide to the Peat Mosses of Boreal North America, by Cyrus B. McQueen. Hanover, N.H., and London: University Press of New England, 1990.

A Focus on Peatlands and Peat Mosses, by Howard Crum. Ann Arbor: University of Michigan Press, 1992.

Peatlands, by P. D. Moore and D. J. Bellamy. New York: Springer-Verlag, 1974.

Olive-sided flycatcher

Hare's-tail cottongrass

Calcareous Fen

A saturated meadow,
      Sun-shaped and jewel-small,
A circle scarcely wider
      Than the trees around were tall;
Where winds were quite excluded,
      And the air was stifling sweet,
With the breath of many flowers,—
      A temple of the heat

ROBERT FROST
"Rose Pogonias"

Fens, like bogs, are open peatlands: wetland ecosystems in which flushing and decomposition of vegetation are slow enough to permit the accumulation of the moist organic soil called peat. Peatlands vary widely in crucial factors such as chemistry, climate, and topography, with the result that "poor" fens and "rich" bogs can be quite similar and contain many species in common. Nevertheless, classic fens and bogs can be readily distinguished by their "field marks."

The clearest distinction between a bog and a fen—one that can create striking ecological differences between these two types of wetland—is that the former is a closed system fed by precipitation with neither inlet nor outlet, while a fen is fed and flushed by some form of running water. Moving water tends to dilute the extreme acidity that builds up in a true bog and that limits the number of species that can tolerate such harsh conditions. It also carries oxygen and minerals into the system, providing support for a richer community of plants and animals than nutrient-poor bogs permit. The water flow also tends to create structural differences. Bogs are typically ponds partially or completely covered with a mat of sphagnum moss inhabited to varying degrees by characteristic shrubs and herbs. Fens more resemble wet meadows dominated by sedges with an uneven surface of hummocks and depressions, dotted with shrubs and small trees and often traversed by one or more meandering rivulets.

Calcareous fens are perhaps the most interesting variations on the peatland theme because they

support unique communities of *calciphilic,* or lime-loving, plants. These fens are concentrated where there is a limestone component to the bedrock. The calcium becomes dissolved in streams or groundwater that flow through the wetland. In New England such places are typically filled-in lake basins with substantial but slowly drained peat accumulations, called basin, or level, fens, or are smaller formations such as seepage fens and hillside fens, where the drainage is faster and the peat accumulations are much less. Weatherbee and Crow (1992) distinguish several different fen communities including forested fens, shrub fens, sloping graminoid fens, lake basin graminoid fens, and calcareous seeps. Calcareous fens are rare and localized in New England and often contain spectacular concentrations of rarities, mainly northern plants that "migrated" south during the last glacial advance and have managed to survive in the relatively cold climate of the fens. In Massachusetts the most diverse and interesting fens are in Berkshire County. Bostonians may wish to note that the present "Fenway" was originally tidal freshwater marsh, not the peatland community described above.

Perhaps the best way to appreciate the special quality of a fen is to come upon one by accident as you are following a stream through a woodland. You see light ahead, you emerge through the last screen of vegetation, and there before you, threaded by the stream, is a bright, enclosed meadow, very different in mood from the rather dark, boreal solemnity that seems to lie on a true bog like ground fog. If you make such a discovery in June, your fen could be ablaze with pink orchids. Look among the sedges for the tiny eastern

Woolly bear caterpillar

red damselfly and for sedge and sphagnum sprites (damselflies) and, in southwestern Massachusetts, for Dion's skipper, one of New England's rarest butterflies. Later in the summer in the best calcareous fens the elegant grass of Parnassus, with its green-veined flowers, blooms in profusion, and in fall pom-poms of cottongrass give the fenscape a decidedly jolly aspect.

To become a serious fensperson, you must sooner or later come to terms with the dominant vegetation of the fens, the infinitely inscrutable sedges. These grasslike plants—and especially the notorious genus *Carex*—are the confusing fall warblers of the plant world, only much worse. As with little brown birds, the human approach to sedges varies. Some choose to ignore identification problems altogether and simply appreciate them in their gracile generality. Others look closely and begin to notice how exquisite many species become on close inspection. And from there it is but a short step to full addiction, leading to endless nights poring over dichotomous keys (translated from the Chinese by Latin scholars) and trying to tease subtle distinctions from pen drawings of "achenes" and "perigynia" that change shape on the page as you stare at them.

INDICATOR SPECIES

❧ TREES: American larch, black ash, red maple, bur oak (SSC). ❧ SHRUBS: Bog rosemary, Labrador tea, and other bog species in more acid sites, sweetgale, alderleaf buckthorn, swamp red currant (SSC), smooth gooseberry (SSC), northern honeysuckle, black chokeberry, shrubby cinquefoil, poison sumac, fen birch (ST). Bog, autumn, hoary, and silky willows. ❧ GRASSES AND SEDGES: Spiked muhly, pendulous bullsedge (SSC), hardstem bullsedge, capillary beaksedge (SE), white beaksedge, twigsedge, foxtail sedge (ST), creeping sedge (SE), dioecious sedge (ST), fen sedge (SSC), aquatic sedge, inland prickly-sedge, chestnut-colored sedge (SE), American sedge, porcupine sedge, yellow sedge, slender woolly-fruited sedge, broad woolly-fruited sedge, delicate sedge, golden sedge, meadow sedge, slender cottongrass (ST), fen cottongrass. ❧ WILD-FLOWERS: Nodding ladies'-tresses, hooded ladies'-tresses (SE), white adder's-mouth (ST), fen orchis, rose pogonia, grass pink, dragon's mouth (ST). The last three orchids occur in more acid, boglike fens, growing in sphagnum. Grass of Parnassus, slender blue-eyed grass (O, ST), three-leaved Solomon's-seal, flat-leaved bladderwort, spatulate-leaved sundew, fen cuckoo-flower (ST), buckbean, brook lobelia, marsh cinquefoil, water avens, tufted loosestrife, northern bedstraw (SE), fen bedstraw (SSC), fringed gentian, swamp thistle, sweet coltsfoot (ST), bog goldenrod, rough-leaved goldenrod. Small yellow lady's-slipper (SE), showy lady's-slipper (SSC), pink pyrola (SE), and hemlock parsley (SSC) are associated with forested fens in the Berkshires. ❧ BUTTERFLIES: Dion's skipper (O, SE) is restricted in Massachusetts to calcareous fens in south Berkshire County; the caterpillar feeds on sedges. Other sedge-feeding butterflies include eyed brown, Appalachian brown, mulberry wing, black dash, dun skipper, and two-spotted skipper. Henry's elfin, a very local, early flying butterfly, is associated with some eastern fens; its caterpillar feeds on wetland hollies and blueberries. ❧ DRAGONFLIES: The odonate fauna of fens has not been well differentiated from that of bogs. Eastern red damsel and sedge and sphagnum sprites prefer sedgy wetlands, including fens. ❧ REPTILES: Bog turtle (O, SE), our rarest turtle, reaches the northern extremity of its range in south Berkshire County. ❧ OTHER VERTEBRATES that may occur in peatlands but are by no means restricted to fens are listed under BOG.

DISTRIBUTION

Calcareous fens occur globally wherever peatland conditions combine with alkaline mineral conditions. In North America they occur mainly in a broad swath of limestone deposits that runs from southeastern Saskatchewan south and east through the Great Lakes states and stops in southeastern New England. In Massachusetts true calcareous fens are restricted to areas of south Berkshire County where the source groundwater flows over underlying bedrock of marble or dolomite. "Poor" fens, those with a lower pH, occur elsewhere in the state in association with calcium deposits.

KEY
O=Obligate
SE=State Endangered
ST=State Threatened
SSC=State Special Concern

Arethusa or dragon's mouth orchid

According to the Massachusetts Natural Heritage and Endangered Species Program, there are fewer than 50 calcareous fens in Massachusetts; perhaps fewer than 20 are of the highest quality. The largest of these is 50 acres but most are small to very small—a few acres or small patches associated with limy seeps.

CONSERVATION STATUS

Calcareous fens have always been rare and of limited extent in Massachusetts. Because their uniqueness depends on variables such as water chemistry, available nutrients, and water levels, they are also unusually susceptible to degradation, and their conservation depends on maintaining the right balance of natural conditions. When this balance is disturbed, fens become particularly susceptible to invasion by aggressive, mainly alien species such as purple loosestrife, European buckthorn, common barberry, Phragmites, and cattail. Fens also tend to be successional communities that over time will support more woody vegetation, to the detriment of rare wildflowers and sedges. Effective conservation therefore may require a dual strategy. On the one hand it is essential to prevent destructive practices such as filling, ditching, and flooding, which alter the delicate water regime, and the introduction of livestock, which brings an excess of nutrients to the system. On the other hand it may be desirable in selected cases to remove invading trees or to undertake careful mowing or even light grazing. The good news is that Massachusetts currently has some of the most diverse and pristine calcareous fens in New England. The bad news is that the state's Natural Heritage and Endangered Species Program considers them the fourth most threatened natural community in the Commonwealth.

PLACES TO VISIT

Because of the rarity and fragility of calcareous fens and the sad fact that unscrupulous collectors have been known to plunder their biological treasures, the Natural Heritage Program discourages publicizing exact locations in most cases.

Kampoosa Fen, Stockbridge. Town of Stockbridge (Area of Critical Environmental Concern).

South Mountain State Forest, Pittsfield. Massachusetts Department of Environmental Management. Arrange visit through DEM Berkshires headquarters in Pittsfield.

FURTHER READING

Bogs of the Northeast, by Charles W. Johnson. Hanover, N.H., and London: University Press of New England, 1985. Includes comprehensive treatment of fens.

"Calcareous Fens of Western New England and Adjacent New York State," by Glenn Motzkin. *Rhodora,* 96 (1994), no. 885, pp 46–68.

Ecosystems of the World. Mires: Swamp, Bog, Fen and Moor, edited by A. P. J. Gore. Parts 4A and 4B. Amsterdam: Elsevier, 1983.

"The Natural Plant Communities of Berkshire County, Massachusetts," by Pamela B. Weatherbee and Garrett E. Crow. *Rhodora,* 94 (1992), no. 878, pp. 203–206.

Peatlands, by P. D. Moore and D. J. Bellamy, New York: Springer-Verlag, 1974.

Spruce-Fir Forest

It was a sweet smelling wood, but very dense, damp, dark and silent except for the occasional odd sounds of invisible birds. The only paths were the random tracks made by wild animals, and the children soon lost their way.

A Tale of the Tontlawald (German folktale)

The spruce-fir forest community, also known as mesic northern conifer forest, boreal forest, or taiga, has several distinctive variations in New England: the fog-bound coastal forests of Maine, dominated by white and red spruce; the mountain, or Appalachian, spruce-fir forest, which crowns the highest, coldest, dampest Massachusetts peaks and ridges and is dominated by red spruce and balsam fir; wetland forests growing on or around peatlands (see BOG), consisting mainly of black spruce and American larch; and so-called *inverted flats*, where red spruce and/or balsam fir sometimes occur below deciduous forest on wet, nutrient-poor soils along lakes and streams. The last three reach Massachusetts as vestiges of the great boreal forests that are typical in northern New England but colonize only the harshest peaks, valleys, and wetlands of the Commonwealth.

Typically, the interior of the spruce-fir community is dense and dark. Its multispired silhouette is distinctive and stately, with mature trees regularly reaching 60 to 70 feet in the best growing conditions. In the highest Massachusetts sites, such as the summit of Mount Greylock, bare rock, high winds, heavy icing in winter, and fires have resulted in a more open community of short trees averaging 20 to 25 feet in height. The trees get smaller still as they ascend higher mountains, finally acquiring the stunted appearance called *krummholz* ("crooked wood" in German), just below the alpine tundra zone on the loftiest New England peaks. Openings in the forest are typically filled with low shrubs or a thick bed of mosses and lichens.

Spruce-fir forest is a moist forest, its distribution on mountaintops conforming closely with the lowest elevations of persistent cloud cover. As the forest ages, high winds knock down bands of the oldest, tallest trees, opening strips of ground to sunlight and seedling germination, a process called *wave regeneration*.

The spruce-fir forest is the deep, dark woods of fairy tales, where hungry wolves wait in the shadows and the old witch keeps a light burning in her cottage to attract wayward children. It is often dank and cold and quite capable of inspiring gloom. On the other hand, it smells wonderful (balsam fir), and it is often popping with interesting wildlife. There is so little spruce-fir forest in

186

Blackpoll warbler

Massachusetts that places like the summit of Mount Greylock and a few forest-fringed bogs acquire a flavor of the exotic, tiny boreal islands in the clouds or in cold basins where one can hear the song of the blackpoll warbler and other voices of the north woods.

INDICATOR SPECIES

Only characteristic boreal species that reach Massachusetts are listed here. ❧ TREES: Red spruce and balsam fir are dominants in uplands; associated species may include eastern hemlock, yellow birch, and heartleaf paper birch. Black spruce and American larch are peatland dominants. ❧ SHRUBS: Northern (or showy) mountain ash (SE), American mountain ash, mountain maple, red elderberry, Bartram's shadbush (ST), mountain holly, velvetleaf blueberry, lowbush blueberry, skunk currant, sheep laurel, Labrador tea, twinflower, creeping snowberry. ❧ WILDFLOWERS: Bunchberry, Canada mayflower, goldthread, starflower, bluebead lily, wild sarsaparilla, common wood sorrel, painted trillium, pink lady's-slipper, large-leaved goldenrod (ST). ❧ SEDGES: Black-fruited wood rush, subspecies *melanocarpa* (SE). ❧ FERNS AND CLUBMOSSES: Evergreen wood fern, mountain wood fern (both varieties of spinulose wood fern), shining clubmoss. ❧ MOSSES AND LICHENS: Feather mosses, broom mosses, reindeer lichen. ❧ BUTTERFLIES: Green comma (has occurred historically in Massachusetts but there are no recent records; caterpillar feeds on willow, birch, alder, and blueberry), white admiral (birch), pink-edged sulphur (has occurred, probably as a vagrant; velvet-leaf blueberry). ❧ DRAGONFLIES: Lake emerald (sole Massachusetts record is from the summit of Mount Greylock; other boreal species in this genus (*Somatochlora*) fly in spruce-fir glades). ❧ BIRDS: There are summer records of yellow-bellied flycatcher (O) at the summit of Mount Greylock, but

KEY
O=Obligate
SE=State Endangered
ST=State Threatened

nesting has never been confirmed. Bicknell's thrush (O) nested at the summit of Mount Greylock until 1972 but no longer breeds in the state. Other typical birds are red-breasted nuthatch, Swainson's thrush (O), golden-crowned kinglet (O), ruby-crowned kinglet (O; very rare breeder in Massachusetts), magnolia warbler (O), yellow-rumped warbler, Blackburnian warbler, blackpoll warbler (O, SSC), mourning warbler, white-throated sparrow (O), dark-eyed junco (O), purple finch. ❧ MAMMALS: Long-tailed shrew, snowshoe hare, deer mouse, red-backed vole, porcupine. The pine marten and lynx have been extirpated from Massachusetts.

DISTRIBUTION

Spruce-fir forest is one of the world's most extensive biomes, covering millions of acres and virtually encircling the globe above 60 degrees north latitude. It also occurs on mountain peaks as far south as Guatemala, in areas high enough to produce a boreal climate. It is the characteristic forest type in the higher elevations of northern New England and along the coast of Maine from about latitude 45 degrees north. In Massachusetts it occurs in isolated patches where the climate is cold and damp, such as the north-facing slopes of the Berkshire Plateau and at the highest elevations such as the summits of Mounts Greylock, Wachusett, and Watatic.

EXTENT IN MASSACHUSETTS

Dickson and McAfee (1988) give a total of 37,000 acres of spruce-fir (red spruce) forest in Massachusetts. However, most of this acreage is timber plantation, and the area of natural spruce-fir forest is much smaller, a few thousand acres at most.

CONSERVATION STATUS

Spruce and fir are mainstays of the commercial timber and wood pulp industries, but natural Massachusetts stands are so small and of such poor quality that they are in little danger of commercial exploitation. Mountain peaks are irresistible to tourists and therefore tend to acquire parking lots, lodges, towers, and telescopes

Mountain ash

(Mount Greylock). They are also fine places to begin ski runs (Mount Wachusett). These factors have caused some destruction and degradation of spruce-fir forest in Massachusetts. Because spruces and firs have shallow roots and are mechanically weak, they are particularly susceptible to landslides, windstorms, and the fate of neighboring trees. And, according to Cogbill (1987), "The montane spruce-fir forests do not replace themselves after major fires as do the boreal forests to the north." The spruce-fir community also seems to be undergoing some form of semi-natural decline in New England, possibly related to acid precipitation, insect pests, fungal disease, and/or climate fluctuation. Some of the changes in the boreal bird populations on the summit of Mount Greylock may be due to the natural maturing of the forest and consequent loss of more open scrubby nesting habitat.

PLACES TO VISIT

Summit of Mount Greylock, Berkshire County; access from North Adams and Lanesboro. Massachusetts Department of Environmental Management (DEM). At the summit, Bascomb Lodge, open April to October, is operated by the Appalachian Mountain Club.

Summit of Mount Wachusett, Princeton. DEM.
Clarksburg State Park, Clarksburg. DEM.

FURTHER READING

"Berkshire Plateau Vegetation, Massachusetts," by Frank E. Egler. *Ecological Monographs* 10 (April 1940), pp. 145–192.

The Boreal Ecosystem, by J. A. Larsen. New York: Academic Press, 1980.

"The Boreal Forests of New England," by Charles V. Cogbill. *Wild Flower Notes* (New England Wildflower Society, Framingham), Fall–Winter, 1987.

North Woods, by Peter Marchand. Boston: Appalachian Mountain Club, 1987.

"Spruce Fir Forests of the Coast of Maine," by Ronald B. Davis. *Ecological Monographs,* 36 (1985), pp. 79–94.

Forest Statistics for Massachusetts—1972 and 1985, by David R. Dickson and Carol L. McAfee. Broomall, Pa.: Northeastern Forest Experiment Station, USDA Forest Service, 1988.

Wildlife Sanctuaries of the Massachusetts Audubon Society

Limestone cobble,
Pleasant Valley Wildlife Sanctuary

Blue Hills Trailside Museum

1904 CANTON AVENUE, MILTON, MA 02186
617-333-0690

The Blue Hills Trailside Museum is the interpretive center for the Metropolitan District Commission's Blue Hills Reservation, a 6,500-acre woodland 10 miles south of Boston that includes Great Blue Hill, Ponkapoag and Houghton's ponds, Ponkapoag Bog, and the Fowl Meadow Reservation (the Neponset River and its floodplain) and offers over 200 miles of trails. The reservation contains a variety of natural communities and rare species of statewide significance. One hundred and seventy species of birds have been recorded on the reservation; the 89 species that nest include the rare worm-eating warbler and a thriving population of turkey vultures. Ponkapoag Bog has the most diverse community of dragonflies and damselflies—over 80 species—recorded anywhere in New England, including the globally endangered banded bog skimmer. The oak-conifer forest harbors several of the Commonwealth's rarest reptiles, and the 22 rocky summits are home to the Hentz's red-bellied tiger beetle.

The Blue Hills are the highest coastal range in the East. At 635 feet, Great Blue Hill, a half-hour climb from the Trailside Museum, affords a commanding view of nearby Boston and its environs and, on a clear day, of Mount Monadnock, 70 miles away.

DIRECTIONS

FROM ROUTE 128 (I-95): Take Exit 2D (Milton) onto Route 138 north. Trailside Museum is 1 mile north of the intersection of Routes 128 and 138, located on the right. The parking lot is on the right before the museum, at the foot of the Great Blue Hill ski slopes.

Broadmoor Wildlife Sanctuary

280 ELIOT STREET (ROUTE 16)
SOUTH NATICK, MA 01760
508-655-2296 OR 617-235-3929

Broadmoor's more than 600 acres border 2 miles of the Charles River in Natick and Sherborn and preserve an area rich in human and natural history. There is a great diversity of habitats: rivers, streams, a former mill pond, oak and pine woodlands, grasslands, and over 250 acres of wooded swamp and marshland. This diversity has attracted 148 bird species, 59 of which nest on the sanctuary. Ruffed grouse, scarlet tanagers, and broad-winged hawks nest in the woods; green-backed herons, wood ducks, and Virginia rails are seen in the marsh, and eastern bluebirds, eastern kingbirds, and swallows are common in the fields.

A striking feature of this landscape is a cluster of vernal pools, which support unique assemblages of invertebrate life and provide breeding habitat for two species of mole salamanders, the (yellow) spotted and the rare blue-spotted.

Among the trail highlights are the Mill Pond/Marsh Trail boardwalk, from which one can see minks, muskrats, herons, ducks, red-winged blackbirds, and dragonflies and in the spring, partridgeberry, pink lady's-slippers, and fringed polygala. The Indian Brook Trail skirts a 5-acre field where indigo buntings, eastern kingbirds, and blue-winged warblers gather in the spring. In the fields watch for red foxes, eastern cottontails, and meadow voles. Other trails run through pine woods, a blueberry swamp with deciduous conifers, and a donut-shaped glacial deposit covered with oak.

DIRECTIONS

FROM ROUTE 16: The sanctuary is on Route 16, 1.8 miles west of South Natick center, on the left. ∾ FROM THE MASSACHUSETTS TURNPIKE: Take I-95 (Route 128) south and exit at Route 9 west. Take Route 9 west to Route 16 in Wellesley. Follow Route 16 west through Wellesley and through South Natick center. The sanctuary is on Route 16 approximately 2 miles beyond South Natick center, on the left.

Drumlin Farm Education Center and Wildlife Sanctuary

SOUTH GREAT ROAD, LINCOLN, MA 01773
617-259-9807

The increasingly rare partnership between land, animals, and people that forms a farm community in the northeastern United States is the focus of Drumlin Farm Education Center, 232 acres of pastures, fields, woodlands, and ponds divided between a working organic farm and a wildlife sanctuary.

The vernal pools, open fields, and woodlands of the

farm provide habitat for distinctive wildlife species such as mole salamanders, bobolinks, and even an uncommon mushroom known as the stinky squid.

Trail highlights include enclosures housing injured or orphaned birds native to New England such as hawks, owls, wild turkeys, and a pair of golden eagles; a burrowing-animal building, where red foxes, woodchucks, striped skunks, and other New England mammals that spend time underground can be seen; and the ever-popular farmyard trail, along which urban visitors can see farm animals, farm machinery, barns, and an organic garden. Baby animals are most abundant in the spring but can be seen at other times of the year as well.

One hundred forty-one bird species have been recorded at the sanctuary, 51 of which are known to nest. Along Boyce Field Trail, tree swallows, bobolinks, Carolina wrens, ring-necked pheasants, killdeer, eastern meadowlarks, and various warblers can all be seen and heard; the woodchucks, red foxes, white-tailed deer, and coyotes that also inhabit these fields show that it is possible to farm and share the land with diverse wildlife. The drumlin itself, the dome-shaped hill left behind by a melting glacier 15,000 years ago, commands a wide view of the surrounding landscape, and is a good spot to watch for soaring hawks and turkey vultures. Eastern bluebirds, indigo buntings, and brown thrashers nest in the dense shrubbery in summer.

DIRECTIONS

BY PUBLIC TRANSPORTATION: The sanctuary is .7 miles from the Lincoln train station. From the station, proceed south on Lincoln Road to Route 117; then turn left. Drumlin Farm is on the right, .2 miles east on Route 117. ∾ FROM I-95 (ROUTE 128): Take Exit 26 (Waltham/Weston) and head toward Waltham on Route 20. After the lights, take the first left; then, at the next set of lights, take a left onto Route 117. Drumlin Farm is 4.5 miles up the road on the left. ∾ FROM ROUTE 117 EAST: The sanctuary is on Route 117, .5 miles east of the intersection of Routes 126 and 117.

Habitat Education Center and Wildlife Sanctuary
10 JUNIPER ROAD, BELMONT, MA 02178
617-489-5050

This sanctuary in the Belmont hills west of Boston was created from the 1994 merger of the Habitat Institute for the Environment, a nonprofit environmental education center, and the Society's Highland Farm property. The property combines old fields and woods with landscape plantings that include the work of four well-known landscape designers: Mary Cunningham,

Eastern kingbirds

William Curtis, and the Olmsted brothers, famous for their work on Boston's Emerald Necklace. In the Highland Farm section, once part of a dairy farm, abandoned pasture slopes down to a series of vernal pools in a red maple swamp and a seasonal brook and bog, as well as a small knoll crowned with white pine and oak.

A single trail meanders through the sanctuary past the greenhouse, along the edge of a meadow where blueberry bushes, goldenrod, and milkweed grow, and through an abandoned apple orchard where yellow-bellied sapsuckers visit the old trees. Sassafras, red-osier dogwood, and American larch also grow nearby. Habitat's bird list totals 139 species of which 54 are breeding. Around a pond, constructed in 1947, watch for green herons, belted kingfishers, eastern kingbirds, eastern phoebes, and vireos. Smooth alder, common spicebush, highbush blueberry, sweet pepperbush, and swamp azalea as well as aggressive alien species such as alderleaf buckthorn and Oriental bittersweet also cover much of the terrain. Along the edges of the pond, notice the arrowwood, sweet flag, and groundnut, and look for crayfish, painted turtles, and bullfrogs.

DIRECTIONS

BY PUBLIC TRANSPORTATION: Take Bus #74 from Harvard Square to Belmont center. The bus stops at the corner of Leonard Street and Alexander Avenue. Walk up Leonard Street (the main street of Belmont center) and cross over Route 60/Pleasant Street. Leonard Street then becomes Clifton Street. Take the first left off Clifton, onto Fletcher Road; then take the first left onto Juniper Road. The sanctuary is a quarter-mile up on the right. ∾ FROM I-95 (ROUTE 128): Exit onto Route 2 east and proceed almost 4 miles. Exit at Pleasant Street (Route 60), Belmont, and take a right at the end of the ramp. At the second set of lights, take a right onto Clifton Street; then take the first left onto Fletcher Road. From Fletcher Road, take the next left onto Juniper Road. The sanctuary is a quarter-mile up on the right. ∾ FROM BOSTON/CAMBRIDGE: Take Route 2 west to the Route 60 exit and take a left at the end of the ramp. At the third set of lights, take a right onto Clifton Street; then follow the remaining directions from Clifton Street, above.

Waseeka Wildlife Sanctuary
CLINTON STREET
HOPKINTON, MA 01748
617-259-9506, EXT. 1601
(EASTERN SANCTUARIES PROPERTY OFFICE)

This 219-acre sanctuary was once farmland, a part of the former Waseeka Farm. (Waseeka is a Native American term for "fertile plain.") Forest has now reclaimed the land, but reminders of its early-twentieth-century agricultural past remain. The sanctuary comprises 86 acres of mixed hardwood forest, 42 acres of mixed forest with pitch pine and large white pines, several acres of abandoned orchards, and a large wildlife pond.

A mile-long trail loops around the property, winding through a forest of oak and white pine before emerging alongside the wildlife pond, whose numerous small islands are covered with swamp azalea and winterberry. Clumps of cattail and masses of water lily cover the pond's surface, interspersed with tree stumps and dead trees, several of which serve as nesting sites for great blue herons. Snapping and painted turtles swim in the pond's depths, and on a warm May day hundreds of painted turtles can be spotted sunning themselves on logs scattered across the pond.

Beyond the pond the trail enters an older forest of larger trees, more maple, and fewer oak, as well as nine huge white pines with trunks 32 to 34 inches in diameter—among the biggest in the area. Yellow stargrass, jack-in-the-pulpit, ground cedar, sassafras, and cinnamon and royal ferns carpet the ground.

DIRECTIONS
FROM I-495: Take Exit 21A (West Main Street, Hopkinton) to Hopkinton center. Merge with Route 135 and continue on Route 135 past Weston Nurseries on the left. Immediately after the nursery, turn right on Clinton Street. Go 2.1 miles, and Waseeka is on the left; look for the Massachusetts Audubon sign and a green gate.

SOUTH SHORE

Allens Pond Wildlife Sanctuary
OFF THE BEACH ROAD AT HORSENECK BEACH
DARTMOUTH, MA 02714
617-259-9506, EXT. 1601
(EASTERN SANCTUARIES PROPERTY OFFICE)

The 188-acre Allens Pond Sanctuary protects an exceptionally beautiful and little-visited stretch of the Commonwealth's spectacular coast. The barrier beach, salt marsh, salt pond, and open heathlands found at Allens Pond are critical to the survival of numerous breeding birds, including the globally threatened piping plover, state-listed tern species, and uncommon species of sparrow. Of the sparrow species, the saltmarsh sharp-tailed and the seaside breed exclusively in this salt-marsh habitat. Allens Pond has 25 to 30 pairs of seaside sparrows, possibly the largest population in New England. Other shorebirds and waterbirds use the rich habitats of this coastal salt pond as a feeding ground. The dynamic cobble beach supports a unique community of plants as well, highlighted by the striking magenta of the rare northern blazing star. Nearby in the Westport River are 5-acre Gunning Island and 3-acre Little Pine, which the Society owns and manages. Both islands are osprey nesting sites, and therefore visitation is discouraged from late spring to fall.

Access to the sanctuary is by a path near the beach. At Allens Pond and its nearby salt-marsh meadows, migrating tree swallows fill up on blue-gray bayberries in the fall before continuing their journey. Besides the bayberry, look for beach rose, cow parsnip, seaside goldenrod, beach pea, northern blazing star, and numerous other plants associated with coastal areas, and watch out for poison ivy. Ducks, great blue and green herons, great and snowy egrets, greater and lesser yellowlegs, willets, and glossy ibis are often seen in the marsh, while ospreys and northern harriers patrol the skies.

DIRECTIONS
FROM I-195: Take Exit 10, near Fall River, and follow Route 88 south to its end. Turn left and follow the beach road until it turns left, away from the beach. The property is on the right. Look for the Massachusetts Audubon Society sign.

Daniel Webster Wildlife Sanctuary
WINSLOW CEMETERY ROAD
MARSHFIELD, MA 02050
617-837-9400
(SOUTH SHORE REGIONAL CENTER)

The exceptional wildlife habitat in the 476-acre Daniel Webster sanctuary includes hayfields and meadows, waterways that empty into nearby Green Harbor Basin, and tidal gates that hold back the ocean from the shallow, slow-moving Green Harbor River. Most of the land is a polder: low land reclaimed from the sea and protected by a dike, as in the Netherlands. The juxtaposition of two tidal rivers, the ocean, and over 350 acres of open grassland habitat with nearby wetlands is perfect for migrating birds and for raptors requiring winter feeding grounds. The abundant small-mammal populations of the grasslands attract northern harriers, rough-legged hawks, and endangered short-eared owls to the sanctuary each winter. As low areas in the grasslands flood each spring, thousands of migrating ducks stop over at the sanctuary to rest and feed. Many birds in the sandpiper family also use the grasslands and wetlands, including yellowlegs, common snipe, least sandpipers, and some rare species such as Baird's and buff-breasted sandpipers and ruffs. Total bird species recorded: 194; total nesting species: 71. Larger mammals include coyotes, red foxes, opossums, raccoons, and cottontails in the meadows and woods, and minks, muskrats, and weasels in the marsh pond and wetlands.

DIRECTIONS

FROM ROUTE 3: Take Exit 12 (Marshfield) and go east on Route 139 to Marshfield center. Go through two sets of lights. After the second set of lights, turn right onto Webster Street. Follow Webster Street 1.4 miles to Winslow Cemetery Road on the left. The sanctuary is at the end of Winslow Cemetery Road.

Moose Hill Wildlife Sanctuary
293 MOOSE HILL STREET
SHARON, MA 02067
617-784-5691

At first glance Moose Hill appears to be an unbroken tract of oak forest, but at least a half dozen uncommon natural communities, representing influences from both northern and southern regions of New England, are found in this 2,000-acre sanctuary. These include examples of Atlantic white cedar swamp, level bogs, pitch pine–scrub oak barrens, and ridgetop grasslands on a 600-foot granite bluff, the second highest point between the Narragansett Basin and Massachusetts Bay. Veins of richer soils and bedrock influence the composition of vegetation as well. Unusual wildflowers and grasses can be found in the boulder-field woodland below the summit of Moose Hill. Fen species, such as bog rosemary, inhabit the edges of Wolomolopoag Pond. The different types of uplands and wetlands provide habitat for a great variety of mammals, resident and migrant birds, reptiles, and amphibians; over 560 species of native plants have been found to date.

A 7-mile section of the Appalachian Mountain Club's Warner Trail, which meanders for 30 miles from Canton, Massachusetts, to Diamond Hill State Park in Cumberland, Rhode Island, runs north to south through Moose Hill Wildlife Sanctuary. It passes by a swamp and then cuts through forest before leading to the summit of Moose Hill. Look for oak, hickory, maple, birch, and cherry along the trail, as well as white pine and hemlock and an understory of witch hazel, dogwood, blueberry, and maple-leaf viburnum. In drier locations, oak and hickory dominate; in wetter locations, red maple and hemlock flourish. Seventy-four bird species nest at Moose Hill, including wood thrushes, red-eyed vireos, Baltimore orioles, and scarlet tanagers in the forest; total bird species recorded: 164.

Another highlight at the sanctuary is a 700-foot boardwalk through a red maple swamp. Wild grape, sedges, and various ferns, as well as skunk cabbage, highbush blueberry, spicebush, and winterberry thrive here. As you walk the boardwalk, you may, in the proper season, see muskrats, turtles, and nesting common yellowthroats.

Among the many other trails is one that loops to numerous vernal pools, where spotted salamanders can be seen in the spring, and continues for 3 miles through a varied landscape of chestnut trees, large rock outcroppings, shady woods with mushrooms and the translucent, waxy Indian pipe, and fields where white-tailed deer, red fox, and coyote are often spotted along the edges.

DIRECTIONS

BY PUBLIC TRANSPORTATION: Moose Hill is less than 2 miles from the Sharon train station. From the station, follow Route 27 north for two blocks, then go left onto Moose Hill Parkway and proceed until you see the entrance on the right. ∾ FROM I-95 SOUTHBOUND: Take Exit 10 and turn left at the end of the exit ramp. Proceed to the next intersection and take a right onto Route 27 north toward Walpole. Follow Route 27 for approximately .5 miles. Turn left onto Moose Hill Street and travel 1.5 miles. The sanctuary is on the left. ∾ FROM I-95 NORTHBOUND: Take Exit 8 and turn right at the end of the ramp onto Main Street. Travel approximately 1 mile and turn left onto Moose Hill Street. Proceed for 1.5 miles; the sanctuary is on the left.

North Hill Marsh Wildlife Sanctuary
MAYFLOWER STREET, DUXBURY, MA 02332
617-837-9400
(SOUTH SHORE REGIONAL CENTER)

An oak and pine forest overlooks a 90-acre pond and North Hill Marsh in this 137-acre sanctuary, which is contiguous to an active cranberry bog and two other ponds. Blueberry, inkberry, and arrowwood grow under the canopy of the oak-pine forest, and the ponds and marsh are rich in aquatic plants and animals, especially migratory birds in the spring and fall. Ring-necked, black, and wood ducks are common, as are buffleheads, herons, and egrets.

North Hill Marsh Wildlife Sanctuary is often said to be the crown jewel of Duxbury's Central Greenbelt, a 1,000-acre linked system of open space and trails that includes Waiting Hill Preserve, Knapp Town Forest, Round Pond, and West Brook, the headwaters of the Back River.

DIRECTIONS

FROM ROUTE 3: Exit Route 3 at the junction with Route 139 (Marshfield). From Marshfield center take Route 3A south for 10.4 miles to Mayflower Street, Duxbury. Turn right onto Mayflower, just after the town hall and cemetery, and go 1.3 miles. The parking lot is on the right. There is limited parking off the street.

North River Wildlife Sanctuary
2000 MAIN STREET, MARSHFIELD, MA 02050
617-837-9400
(SOUTH SHORE REGIONAL CENTER)

The 175-acre North River Wildlife Sanctuary—a mix of habitats including hardwood forest, mixed pine and hardwood, red maple swamp, salt marsh, and open fields—attracts an unusual variety of birds and wildlife. It borders the historic estuarine North River, notable for its spectacular marshes where the river empties into Massachusetts Bay. Here in midsummer shorebirds that have left their nesting grounds on the northern tundra stop en route to South American wintering grounds; in early fall boreal forest warblers flying south stop at the sanctuary for short periods. Predatory merlins and peregrine falcons often shadow the shorebirds along their migratory coastal routes. White-winged, black, and surf scoters can be spotted along the North River in autumn, as well as red-throated and common loons, some of which winter along the marsh and river. Horned and red-necked grebes and an occasional common eider are also seen. Migrating bald eagles often soar about the sanctuary in the fall, and spring brings flocks of warblers, including some of the less common species.

A 1-mile riverside boardwalk trail leads through a red maple swamp, cattail marsh, and salt marsh overlooking the North River. In winter, look for harbor seals in the river. Wild grapes grow in great tangles along parts of the trail—their aroma fills the air on sunny autumn days.

DIRECTIONS

FROM ROUTE 3: Take Exit 12 and go east on Route 139 toward Marshfield. At the first set of traffic lights, turn left onto Route 3A. Go 4.6 miles. The sanctuary entrance is on the right. ❧ FROM ROUTE 3A SOUTH-BOUND: Cross over the North River Bridge at the Scituate-Marshfield town line. The sanctuary is a half mile farther on the left.

Stony Brook Nature Center and Wildlife Sanctuary
108 NORTH STREET, NORFOLK, MA 02056
508-528-3140

Along the northern edges of Stony Brook Wildlife Sanctuary stretch Kingfisher Pond and Teal Marsh, part of an extensive wetland interspersed with fields, meadows, and a hardwood and pine forest. The sanctuary's 241 acres attract 160 species of birds, of which 95 nest. Oak and white pine dominate the woods, but look also for dogwood, red maple, beech, and white birch along the boundaries of the sanctuary.

A 1.3-mile-long self-guided trail winds through the sanctuary, starting in a meadow where swallows and eastern bluebirds make their homes. A boardwalk snakes through Teal Marsh, where turtles, fish, muskrats, and great blue herons live. Near the marsh edge, blueberry and sweet pepperbush grow in profusion and red-winged blackbirds nest in clumps of cattails and tussock sedge, while out in the shallow water tuft grasses and sweet gale sprout on small undulating islands that provide nesting sites for yellow warblers. The boardwalk trail winds through a grove of beech trees to an observation platform on an island overlooking Kingfisher Pond. Arrowhead, pickerel weed, and lilies grow in the aging pond as mud and sediment fill it in.

In the broad reaches of Kingfisher Pond, hummocks of alder and buttonbush nestle along the shoreline and provide breeding cover for mallards, black ducks, and Canada geese. The sanctuary also features a butterfly garden with flowers chosen because they are desirable nectar sources for adult butterflies or have leaves that are food for caterpillars.

DIRECTIONS

BY PUBLIC TRANSPORTATION: Stony Brook is 1.2 miles from the Norfolk train station. Go south on Route 115; then turn right onto North Street (at the sign for the sanctuary). The sanctuary buildings are the first ones on the right. ❧ FROM THE JUNCTION OF ROUTES 1A AND 115 (IN NORFOLK): Go north on Route 115 for 1.5 miles and take the third left onto North

Street. The sanctuary is on the right a few hundred yards up North Street.

NORTH SHORE

Eastern Point Wildlife Sanctuary
EASTERN POINT BOULEVARD
GLOUCESTER, MA 01930
508-887-9264
(IPSWICH RIVER WILDLIFE SANCTUARY)

The wet thickets, oak woodland, open fields, salt marsh, freshwater cattail marsh, and tidal pools that make up the 36-acre Eastern Point Wildlife Sanctuary in East Gloucester attract migratory birds and various species of gulls and cormorants. The tidal flats attract shore-birds in spring, summer, and fall, while loons, grebes, sea ducks, gulls, and alcids appear on the ocean and in the adjacent waters of Gloucester Harbor in winter.

In the fall, monarch butterflies in numbers roost in the trees as they migrate along the coast, heading to the mountain forest in central Mexico where they spend the winter. The sanctuary is a reliable spot for observing the butterflies in October. In the dense, oak-dominated woodland are Solomon's-seal, witch hazel, trout lily, dwarf ginseng, and wood anemone.

DIRECTIONS
FROM ROUTE 128: Follow 128 northward to Exit 9. Continue straight to East Main Street, Gloucester. Proceed on East Main Street for 2 miles; then bear right onto Eastern Point Boulevard. Follow the road to the parking area at the end, near the Coast Guard Lighthouse. The main body of the sanctuary is on the eastern side of Eastern Point Boulevard.

Endicott Wildlife Sanctuary
346 GRAPEVINE ROAD, WENHAM, MA 01984
508-927-1122
(NORTH SHORE CONSERVATION ADVOCACY)

The 36-acre Endicott Sanctuary, a former estate, contains a surprising variety of topography, geological formations, and plant and animal communities. Twenty-six species of butterflies have been observed in the formal garden and the nearby wet meadow. The surrounding terrain was sculpted by glacial activity that produced ledges, now moss-covered, and rocky crags; several vernal pools nestle in among the granite knolls. The woods are mostly white pine, hemlock, and oak, with an understory of winterberry holly, sweet pepperbush, and highbush blueberry. The granite out-

croppings are covered with polypody ferns and rock tripe lichens. Pileated woodpeckers, winter wrens, and Louisiana waterthrushes have all been seen on the sanctuary. Great horned owls have nested in the large white pines, and flying squirrels are active at night.

Visitors may enjoy the butterfly garden during weekday office hours. There are no walking trails.

DIRECTIONS
FROM ROUTE 128: Go north on Route 128 to Exit 17 (Grapevine Road). At the exit, turn right toward Beverly Farms. The center is the third driveway on the right, several hundred yards from the exit.

Ipswich River Wildlife Sanctuary
PERKINS ROW, TOPSFIELD, MA 01983
508-887-9264

Eight miles of the Ipswich River meander through the woods and meadows of the 2,800-acre Ipswich River sanctuary north of Boston. Along its course one finds heavy sections of silver maple in the floodplain forest with an occasional river birch—rare in eastern Massachusetts—interspersed with meadows of reed canary grass and patches of rare sedges. This sanctuary is largely wetlands and includes the 2,500-acre Great Wenham Swamp, the largest wetland on the North Shore. The meadows and marshes of the Great Wenham Swamp provide habitat for such rare birds as the American bittern, rails, and the pied-billed grebe, as well as waterfowl and marsh wrens. In the summer months the wetlands come alive with hatches of dragonflies and damselflies.

Sections of the Great Wenham Swamp are covered with thickets of black ash, while elsewhere the very rare and showy lady's-slipper grows. The open marsh is dotted with islands and blends with meadows, fields, old pine and hardwood forests, a sphagnum bog, and an Atlantic white cedar swamp. The riverine and mixed upland forests harbor a great variety of migrant and resident songbirds. A drumlin, an esker, and a kame—geologic formations that serve as reminders of the last glacier, which sculpted this landscape 15,000 years ago—dot the property.

Over 10 miles of interconnecting trails invite visitors to explore the sanctuary. Averill's Island Trail leads past Hassocky Meadow, where cattails and reed canary-grass grow, to the marsh, where ducks and geese feed, on into a white pine forest. American beech, eastern hemlock, and several species of oak can be seen, as well as a rare stand of native red pine, and red maple and yellow birch. Bunker Meadows Trail leads to a buttonbush shrub swamp where, in early spring and fall, ring-necked ducks, buffleheads, green- and blue-winged teal, and wood ducks congregate. In April, listen for the chorusing of American toads and pickerel frogs, and watch for river otters, painted turtles, tree swallows, and great blue herons at various times throughout the year. Along the Rockery Trail, exotic trees and shrubs are reminders that this was once a private estate.

Two hundred and nine bird species have been recorded at Ipswich River, of which 92 are known to nest.

DIRECTIONS

FROM BOSTON: Take I-95 north and exit onto Route 1 north, to the lights. Follow Route 1 to Route 97; turn right (south) onto Route 97, and take the second left onto Perkins Row. Continue 1 mile to the sanctuary, on the right.

Common nighthawk

Marblehead Neck Wildlife Sanctuary and Nahant Thicket

MARBLEHEAD, MA 01945
508-887-9264
(IPSWICH RIVER WILDLIFE SANCTUARY)

Marblehead Neck, 15.5 acres of dense thickets, woodlands, swamps, ponds, and rocky outcropppings, is a hot spot for birding from April to early June and again from August through October, with both common and uncommon migratory birds stopping by; 235 species have been recorded there, including the great majority of eastern North American songbirds.

Stiff-leaved aster and downy goldenrod grow on rocky outcrops above the main pond, and poison sumac grows in the wooded swamp (note the white berries that distinguish it from the other, harmless, sumacs, which have red berries). A Scotch pine plantation attracts migrants and sometimes shelters saw-whet and long-eared owls in fall and winter. Note the bayberry- and huckleberry-covered heath, created when the original oak forest was cleared by settlers who cut pastures for sheep. The small pond off the main trail attracts migratory birds, including many relatively rare warblers such as hooded, Kentucky, prothonotary, worm-eating, and mourning.

Nahant Thicket is a brushy, wooded 4-acre sanctuary on the southeastern part of Nahant Peninsula. Although it has no ocean frontage, the site is frequented by a variety of migratory birds during fall and spring.

DIRECTIONS

For Marblehead Neck

FROM BOSTON: Go north on Route 1 and exit onto Route 128 north. From Route 128, exit onto Route 114

east (Peabody), and then proceed on Route 114 to Marblehead. Turn right at Ocean Avenue (Route 129) and go through the traffic lights; cross the causeway and bear right at the fork. Continue on Ocean Avenue and take a left onto Risley Road. There is a small parking area.

For Nahant Thicket

FROM BOSTON: Follow Route 1 and exit at Route 129 (Lynn and Swampscott). In Lynn, turn right on Shore Drive, which becomes Nahant Road. Follow Nahant Road across the causeway and almost to the end of Nahant. Turn right onto Wharf Street and take the left fork. The sanctuary is on the right, south of the road.

CENTRAL MASSACHUSETTS

Broad Meadow Brook Wildlife Sanctuary

414 MASSASOIT ROAD, WORCESTER, MA 01604
508-753-6087

Broad Meadow Brook, the largest urban wildlife sanctuary in New England, is located in the city of Worcester on 277 acres of oak woods, old pasture and fields, streams, marsh, and swamp. Mammals that live on the property include red foxes, white-tailed deer, raccoons, eastern cottontails, and mink. The sanctuary has hosted 131 bird species including 54 nesters as well as rarities such as orchard oriole, willow flycatcher, and blue grosbeak. Volunteers working for the Massachusetts Butterfly Atlas Project recorded 66 species in the sanctuary, including the southern hairstreak, listed by the state as rare. This diversity, the highest for any of the Society's sanctuaries, represents over 60 percent of all the butterfly species known for Massachusetts.

The brook for which the sanctuary is named flows through underground culverts in Worcester to emerge in the sanctuary before continuing its course into Dorothy Pond in Millbury and on into the Blackstone River. Watch for minks and muskrats near the brook and note the skunk cabbage, wood anemone, marsh marigold, witch hazel, and Indian poke growing in the moist soil along the bank. Pioneer trees in forest succession such as pin cherry and quaking aspen are dominant near abandoned pastureland, and below oak, hickory, ash, and maple form an established hardwood forest. False Solomon's-seal, smooth Solomon's-seal, jack-in-the-pulpit, nodding trillium, wild geranium, jewelweed, blue-stemmed goldenrod, and white wood aster are found here in season. In the second-growth oak forest on the west side of the brook, look for Indian pipe, blueberry bushes, wintergreen, and Indian cucumber root. Beaked hazelnut shrubs are the first wild blossoms, showing tiny, scarlet blooms in March.

There are four major wetlands within the sanctuary: a Phragmites marsh, a cattail marsh, a sedge marsh, and a wooded swamp. Look for blue flags, sphagnum moss, and arrowwood. Bats, red-winged blackbirds, tree swallows, eastern screech-owls, northern water snakes, and turtles (snapping, painted, and the rare spotted) all frequent the wetlands.

DIRECTIONS

FROM THE MASSACHUSETTS TURNPIKE: Take the Millbury exit and at the bottom of the ramp turn onto Route 122 north. Take the exit for Route 20 west. At the first traffic signal, turn right on Massasoit Road. The sanctuary is .5 miles ahead on the left. ∾ FROM ROUTE 20 HEADING EAST: After the junction of Routes 146 and 20 go another 1.5 miles; take a left onto Massasoit Road. ∾ FROM ROUTE 290: Westbound, take Exit 13 at Vernon Street and turn left. Eastbound, take Exit 13 at Vernon Street and turn right. Proceed up Vernon Street and take a left at the fork onto Winthrop Street. Follow Winthrop past Providence Street and St. Vincent's Hospital. Winthrop Street becomes Heywood Street. Go straight on Heywood to the bottom of the hill and turn right onto Massasoit. The sanctuary is 1.4 miles ahead on the right.

Burncoat Pond Wildlife Sanctuary
OFF ROUTE 9, SPENCER, MA 01562
508-355-4638
(WORCESTER COUNTY PROPERTY OFFICE)

The fields and woods of the 126-acre Burncoat Pond sanctuary are home to flying squirrels, weasels, eastern cottontails, snowshoe hares, red and gray foxes, ring-necked pheasants, ruffed grouse, American woodcocks, and great horned owls. The adjacent fields and old orchards attract numerous butterflies as well as northern orioles, cedar waxwings, and eastern phoebes. At the northern edge of Burncoat Pond, ducks, great blue herons, and migrating ospreys often visit.

The main trail of this sanctuary leads through a large field bordered on the left with old oaks, maples, and an apple orchard. On a warm late-summer day, monarch butterflies glide through the fields, alighting to lay eggs on the milkweed leaves, a favorite food for their larvae. The trail then enters a forest where red oak, hemlock, and mountain laurel grow in profusion, and several large beech trees flourish near the path. Great horned owls nest in the woods and can sometimes be spotted at twilight patrolling for prey. The path ends at a small knoll overlooking a shallow bay of the pond called Audubon Cove. Duckweed and water lily cover large sections of the pond in summer; several small fingers of rocky land jut into the pond and create minibays with shallow waters that attract dragonflies.

DIRECTIONS

FROM ROUTE 9: Traveling west from Leicester toward Spencer, turn left onto Polar Spring Road just after the Spencer town line. Polar Spring Road is a dirt road next to the paved driveway of the Spencer Country Inn. Follow Polar Spring Road a short distance until it stops at sanctuary gate. Parking is limited near the gate.

Cook's Canyon Wildlife Sanctuary
141 SOUTH STREET, BARRE, MA 01005
508-355-4638
(WORCESTER COUNTY PROPERTY OFFICE)

Two parcels, Cook's Canyon (40 acres) and Williams Woods (11 acres), make up this sanctuary. A system of easy trails leads through meadowlands and a mixed forest of hemlock, red pine, oak, maple, and birch that blends into 2 acres of marsh and wooded swamp. The main trail parallels Galloway Brook, which flows below high rock ledges. Farther on the brook is dammed, forming Canyon Pond. In spring, starflowers grow alongside the trail, and hog peanut is also found there—look on the ground for a delicate twining vine with pointed light green leaflets in three's. Canada mayflower, partridgeberry, elderberry, wild yellow flax, and mountain laurel all grow at various spots along the trail. Watch and listen for veeries, eastern wood pewees, scarlet tanagers, and ovenbirds during nesting season. Wood ducks, chimney swifts, indigo buntings, bobolinks, blue-winged warblers, and black-throated green warblers also breed here.

DIRECTIONS

FROM THE MASSACHUSETTS TURNPIKE: Take Exit 8 (Palmer) onto Route 32 north. Follow the road for 24 miles, to the Barre town common. From the common, follow South Street and continue for one-quarter mile; the sanctuary entrance is on the left.

Flat Rock Wildlife Sanctuary

529 ASHBURNHAM HILL ROAD
FITCHBURG, MA 01420
508-355-4638
(WORCESTER COUNTY PROPERTY OFFICE)

Old bridle paths and walking trails wind through a landscape of wooded knolls, open fields, sphagnum wetland, woodland streams, and beech, hemlock, and oak groves on this 360-acre site located not far from downtown Fitchburg. Flat Rock Hill, for which the property is named, is at the top of a 950-foot-high slope with a fine view of the surrounding area and the Nashua River Valley. The trail system is a remnant of the days when the property was crisscrossed by bridle paths, and several of the main trails are underlain with meticulous rock work. One trail climbs a small hill and loops back, passing stands of beech, oak, and hemlock. At the top of the hill, a knoll looks out over an open meadow dotted with red cedar and lowbush blueberry. In autumn, hawks migrating south soar on thermal updrafts nearby. Another trail swings by a red maple swamp bordered in places by swamp azalea. In June, note the mountain laurel in the forest nearby and listen

for wood thrushes, veeries, ovenbirds, and scarlet tanagers, as well as various warblers breeding at the sanctuary.

DIRECTIONS

BY PUBLIC TRANSPORTATION: The sanctuary is 3 miles from the Fitchburg train station. Follow the directions below from Fitchburg center. ∾ FROM ROUTE 2: Exit at Route 31 (eastbound) or 13 (westbound) and proceed to Fitchburg center, the junction of Routes 31, 13, and 2A. Take Route 2A and turn right on West Street. Take the second right onto Ashburnham Hill Road. The sanctuary entrance is 1 mile ahead on the right.

Lincoln Woods Wildlife Sanctuary

226 UNION STREET, LEOMINSTER, MA 01453
508-537-9807

In this 68-acre wildlife sanctuary white pine and mixed deciduous trees dominate the landscape, and a glacial esker ridge, molded into the landscape some 15,000 years ago, runs north and south through its southern half. Six vernal pools nestle in the hollows on either side of the esker. These temporary pools cannot support fish, but for this very reason they contain a rich variety of aquatic invertebrates, both insects and crustaceans, including the rare intricate fairy shrimp. During March and April they also host several species of breeding amphibians such as wood frogs, spring peepers, and spotted salamanders. In midspring wildflowers carpet the meadow, and numerous species are evident. Ambush bugs, eastern garter snakes, indigo buntings, little brown bats, and meadow voles are a few of the meadow species that can be observed.

A parade of bird life passes through the sanctuary during spring and fall migration. Look for warblers, thrushes, and flycatchers in the dense underbrush and forest that surround the vernal pools. The sanctuary also attracts red foxes, opossums, striped skunks, and raccoons. In the fall and spring, look for a variety of hawks that use the sanctuary as a hunting stop on their migratory routes. Great horned owls also hunt in the sanctuary.

DIRECTIONS

FROM ROUTE 2: Exit onto Route 12 south (Exit 31); then follow Route 12 south to Leominster center and go left at the city square. Beyond the intersection of Routes 12 and 117, turn right onto Union Street. The sanctuary is .7 miles down Union Street on the left.

Northern cardinal

Wachusett Meadow Wildlife Sanctuary
113 GOODNOW ROAD, PRINCETON, MA 01541
508-464-2712

Wachusett Meadow, one of the largest Massachusetts Audubon Society sanctuaries, stretches across 1,031 acres of old farmland, meadow, forests, wetlands, and shrub hilltop. The great diversity of the site has attracted over 152 species of birds, including 96 breeding species, among them bobolinks, black-throated blue warblers, American woodcocks, least flycatchers, eastern bluebirds, and hermit thrushes. At 2,000 feet, Mount Wachusett is the highest peak in Massachusetts east of the Connecticut River. Many of the species at Wachusett Meadow reflect the slightly colder niches provided at these higher elevations of Worcester County. Among the two dozen rarities reported for the sanctuary are rock shrews and water shrews.

Mount Wachusett, Little Wachusett, Leominster State Forest, and the sanctuary are part of a large block of preserved contiguous land that allows animals that require larger ranges, such as black bears, fishers, river otters, and eastern coyotes, to take up residence. Fringed and bottle gentian, many species of aster and goldenrod, round-leaved violet, painted trillium, and three-toothed cinquefoil are part of the great diversity of wildflowers that carpet the sanctuary in spring, summer, and fall. White ash, hop hornbeam, yellow and paper birch, hemlock, and white pine are all common, as well as shrubs such as highbush blueberry. The 300-year-old Crocker maple, one of the largest sugar maples in North America, is a well-known feature. The meadows attract a good mix of butterflies, including uncommon species such as the Harris checkerspot, Arctic skipper, pepper-and-salt skipper, and white admiral.

Eleven miles of trails through this sanctuary include a boardwalk through a vast red maple swamp and the Rock Pasture Trail, which leads through secluded sections of mature forest where large hemlocks block the sun and the air is cooler than in the fields and other parts of the woods. A mountain trail climbs uphill toward Mount Wachusett, which is one of the finest hawk-watching points in New England. In late spring and early autumn, the skies fill with up to 3,000 hawks a day riding the thermal updrafts as they follow their migratory routes between Canada and Central and South America. On a clear day the view from Mount Wachusett summit is over 100 miles in all directions.

DIRECTIONS
FROM ROUTE 2: Exit onto Route 31 south and proceed to the center of Princeton, where Route 31 intersects with Route 62. Take Route 62 west .7 miles, and turn right at the Massachusetts Audubon sign onto Goodnow Road. The sanctuary is 1 mile ahead. ❧
FROM THE MASSACHUSETTS TURNPIKE: Exit onto Route 290 north into Worcester, continuing north on Route 190 to Route 140. From Route 140, turn onto Route 62 west and follow it to Princeton. Follow the directions above from Princeton.

BERKSHIRES AND CONNECTICUT RIVER VALLEY

Arcadia Nature Center and Wildlife Sanctuary
127 COMBS ROAD, EASTHAMPTON, MA 01027
413-584-3009

Arcadia lies along the shore of an oxbow of the Connecticut River that was the inspiration for Thomas Cole's painting "The Oxbow." Arcadia's 550 acres include upland forest, floodplain forest, meadows, an ancient marsh, and other wetlands. The oxbow ecosystem at Arcadia is the only one of its kind on the Connecticut River that is still intact and wholly owned by a conservation organization. The Mill River, a slow-flowing, sandy-bottomed tributary of the Connecticut that meanders through portions of the ancient oxbow, is bordered by floodplain forest dominated by silver maple, giving way to a high-terrace floodplain forest containing a large diversity of tree species. A dark summer's night may provide an unforgettable display of fireflies over these terraces. The wetland and aquatic habitats are rich in rare plants, turtles, and especially, mussels and snails.

Five miles of trails allow access to the edges of the marsh, a shrub swamp, and diverse upland habitats. Trees typical of floodplain habitats—cottonwood, silver maple, and sycamore—shade Arcadia's paths.

Hemlock slopes wedged above sand and clay streams are part of a landscape created by the formation and disappearance of ancient Lake Hitchcock nearly 15,000 years ago. Chestnut stumps are a reminder that those great trees were once common here.

Arcadia's floodplain forest is an example of an endangered habitat in Massachusetts. The floodplain and associated wetlands are home for an assortment of bird life including wood ducks, great blue herons, egrets, belted kingfishers, and green herons. Great horned owls are often seen in the nearby woods. Other birds nesting in Arcadia include red-tailed and Cooper's hawks, pileated woodpeckers, American robins, brown thrashers, and scarlet tanagers. Migration brings warblers and waterfowl. Total bird species recorded: 215; nesting: 84. White-tailed deer, black bears, river otters, beavers, muskrats, red foxes, flying squirrels, minks, and an occasional coyote can be seen in the sanctuary.

DIRECTIONS

FROM ROUTE I-91: Take Exit 18 (Northhampton) and proceed south on Route 5. Cross the Easthampton town line, and turn right onto East Street toward Easthampton center. After 1 mile watch for the Massachusetts Audubon sign on the right, at Fort Hill Road. Follow the signs to the sanctuary. ❧ FROM NORTHAMPTON: Take Route 10 south, toward Easthampton. Watch for the Audubon sign on your left at Lovefield Street, before the Easthampton town line. Follow the signs to the sanctuary.

Canoe Meadows Wildlife Sanctuary
HOLMES ROAD, PITTSFIELD, MA 01201
413-637-0320
(PLEASANT VALLEY WILDLIFE SANCTUARY)

The Housatonic River flows along the edge of this 262-acre sanctuary of meadows, wetlands, pine forests, and croplands. The streams, wetlands, and floodplain forest associated with the river support over a dozen rare butterfly, bird, reptile, amphibian, and plant species. The shrubby meadows and hayfields draw bobolinks, tree swallows, yellow warblers, and indigo buntings; the river and its adjacent swamps have wood ducks, black ducks, and green herons; and the woodlands support ruffed grouse, veeries, pine warblers, pileated woodpeckers, and red-breasted nuthatches. The uplands and high open fields are excellent for raptor watching: American kestrels and Cooper's and red-tailed hawks often swoop by. White-tailed deer, raccoons, and red foxes roam the meadows and cornfields, and an occa-

sional black bear may visit for some corn. The rare wood turtle is also found at the sanctuary.

Three miles of trails include the Wolf Pine Trail, named for a broad-spreading white pine left in a farmer's field over a century ago, where it grew to its great bulk in the absence of competition. The trail winds through hemlock woods where red-breasted nuthatches call from the trees and barred and great horned owls sometimes nest. Another interesting route, the Sacred Way Trail, leads from West Pond through meadows. In summer, small, floating green plants called duckweed cover the pond, and painted turtles sun themselves on floating logs. Nearby is an oxbow pond, once the main channel of the Housatonic River, with stands of gray birch and quaking aspen as well as a large black maple—an uncommon species in the Berkshires. Silver maples, box elder, American elm, and sycamore are all typical of river floodplains.

The Housatonic River glides past open fields of meadow sweet, steeplebush, and a variety of sedges, rice cutgrass, reed canary-grass, and Phragmites. In the backwaters, look for tussock sedge, bulrushes, pussy willow, and cattails along with thickets of alder, dogwoods, and viburnums.

DIRECTIONS

BY PUBLIC TRANSPORTATION: Take the Holmes Road bus (#18). In downtown Pittsfield, it can be boarded in front of Newberry's on North Street. The bus runs hourly every day but Sunday. ❧ FROM THE MASSACHUSETTS TURNPIKE: Take Exit 2 (Lee) and travel north on Route 7/20 through Lee, toward Lenox and Pittsfield. Eight miles from the turnpike exit, turn right onto Holmes Road and proceed for 2.7 miles. The sanctuary entrance is on the right. ❧ FROM DOWNTOWN PITTSFIELD: Take East Street east to Elm Street. Turn right off of Elm Street onto Holmes Road and follow the road 1 mile to the sanctuary entrance on the left.

High Ledges Wildlife Sanctuary
OFF PATTEN HILL ROAD
SHELBURNE FALLS, MA 01370
617-259-9506, EXT. 1601
(EASTERN SANCTUARIES PROPERTY OFFICE)

High Ledges sanctuary was named for a series of prominent ledges that overlook the scenic Deerfield River; numerous ledges bisect the sanctuary, including an escarpment and cave that is said to have been occupied by the last pair of wolves in the area. A mixed hemlock and hardwood forest, hemlock ravines, a beaver pond, wooded swamps, old fields and pasture, and seeps and springs create a great mix of habitats. This 500-acre site is especially notable for the rich diversity of its plant life. Orchids and ferns are among the most notable plant groups, with 20 of the former and 30 of the latter recorded. The calcareous schists that outcrop occasionally at this eastern end of the Berkshire Plateau account for much of the diversity. The ledges also support a native red pine stand, an uncommon forest type in Massachusetts. About 105 bird species, including 50 nesters, have been observed at the sanctuary, as well as beavers, coyotes, red foxes, otters, black bears, bobcats, and other mammals.

Several trails cross the property, as well as a well-maintained dirt road to the 1,350-foot-high ledges for which the sanctuary is named. From the top of the cliff there is a sweeping, wide-angle view of the Deerfield River Valley and the village of Shelburne Falls below. The trail leads north from the road, about halfway between the town road and the ledges, along an alder swamp where northern waterthrushes nest. Farther along in the woods, red and painted trilliums are abundant, and many common woodland birds can be seen or heard during breeding season. Yellow-bellied sapsuckers and pileated woodpeckers nest near the trail. Another trail leads through a grove of red pines where the ground is carpeted with pink lady's-slippers in late May and early June. Several patches of yellow lady's-slippers and most of the sanctuary's 10 species of violets can be found along the trail, as well as many species of ferns. Still another trail through this remarkable, diverse property leads to the Gentian Swamp, a small open area with a dense population of pitcher plants (blossoming in early June) and grass of Parnassus (early September), along with turtlehead, Joe Pye weed, and other marsh-loving plants.

DIRECTIONS

FROM ROUTE 2 OR I-91: From the traffic circle in Greenfield (junction of Route 2 and I-91), take Route 2 west for 6 miles. Turn onto Little Mohawk Road on the right (across from Shelburne Center Road). Follow the signs for High Ledges, bearing left twice at forks. Turn right on Patten Hill Road and follow it to the second entrance road on the left and a sign for High Ledges. Follow the dirt entrance road .8 miles, until you reach a small parking area on the grass to the left of a gate. The road to the sanctuary is closed from November to April.

Laughing Brook Education Center and Wildlife Sanctuary
793 MAIN STREET, HAMPDEN, MA 01036
413-566-8034

The children's author Thornton W. Burgess (1874–1965), the creator of Peter Rabbit and his friends, named the stream that runs near his summer home in Hampden Laughing Brook because it reminded him of the sound of children laughing. Today the brook is part of the 354-acre Laughing Brook Wildlife Sanctuary and remains as pristine and sonorous as in Burgess's day. A nearby pond will also be familiar to children as the Smiling Pool from the Old Mother West Wind stories. Laughing Brook's 4½ miles of trails lead from the brook, through an upland deciduous forest, fields, and swamp habitats. Forest types include oak, hickory, beech, and maple. A grove of climax hemlock is part of 175 acres of undisturbed forest that provides excellent opportunities for viewing birds and wildlife. An esker, a long, narrow ridge of glacial debris, borders the brook, where minks and muskrats live.

One hundred thirty-six bird species have been recorded at Laughing Brook, including 66 nesters.

Seven species of rare reptiles and amphibians are present in this sanctuary and those, combined with the more commonly seen species, add up to over 60 percent of the Massachusetts "herptile" species. In the mixed hardwood forests and open fields watch for ovenbirds, barred owls, porcupines, wild turkeys, and coyotes. Rattlesnake plantain grows in certain damp areas.

DIRECTIONS

FROM THE MASSACHUSETTS TURNPIKE: Take Exit 8 (Palmer), and turn right into Palmer center. Follow Route 20 west (toward Springfield) 5.3 miles, turn left onto Main Street, Wilbraham, then follow Main Street for 6.6 miles. At the stop sign, take a left onto Allen Street, go .2 miles, and take a left onto Main Street, Hampden. The sanctuary is two miles ahead on the left. ∾ FROM I-91: Take Exit 4 (northbound) or Exit 2 (southbound) onto Route 83 south, which becomes Sumner Avenue. Follow Sumner Avenue for 8.1 miles (it becomes Allen Street 3.5 miles along the way) and take a left onto Main Street, Hampden. The sanctuary is two miles ahead on the left. ∾ FROM STAFFORD SPRINGS, CONNECTICUT: Take Route 32 north toward Monson. From the Massachusetts state line, go 4.9 miles on Route 32, then take a left in front of the white church and then an immediate right onto High Street. After 2.1 miles on High Street, bear left at the fork onto Upper Hampden Road. The sanctuary is 4.2 miles ahead on the right. ∾ FROM SOMERS, CONNECTICUT: Starting from the intersection of Routes 190/83, take Route 83 north for 2.6 miles, then turn right onto Hampden Road. Follow Hampden Road for 2.6 miles (it becomes Somers Road after the state line), then turn right onto Main Street, Hampden. The sanctuary is two miles ahead on the left.

Pleasant Valley Wildlife Sanctuary
472 WEST MOUNTAIN ROAD, LENOX, MA 01240
413-637-0320

Stretching along the rugged east flank of Lenox Mountain in southern Berkshire County and descending to Yokun Brook before rising again, Pleasant Valley Wildlife Sanctuary offers 1,200 acres of high hills, hardwood forest, meadows, wetlands, and beaver ponds. Seven miles of trails wind through the property, traversing old fields, wooded uplands, and swamps. The hills on the slope of Lenox Mountain feature steep rocky ledges and hemlock gorges, while a limestone cobble in the northern section of the sanctuary supports unusual plant species. Yokun Brook, fed by numerous small brooks, flows through the sanctuary from south to north and is home to a thriving beaver population.

This sanctuary is an important biological reserve for the state, with populations of over 40 rare or uncommon species and a flora that approaches 700 species. The calcareous soils and bedrock of the region account for much of this richness and are indicated by such natural community types as rich mesic forest, dry hickory–hornbeam forest, black ash swamp, sloping fens, and calcareous ledges and cobbles.

The many habitats of the sanctuary have attracted 156 species of birds to date, 89 of which breed at the sanctuary. A hummingbird garden near the office draws ruby-throated hummingbirds in season, and nesting boxes are home to tree swallows. The forest trails lead to areas where wild turkeys, ruffed grouse, pileated woodpeckers, yellow-bellied sapsuckers, winter wrens, least flycatchers, yellow-throated vireos, black-throated blue and black-throated green warblers, and Louisiana waterthrushes can all be spotted. Though seldom seen, black bears also roam these woods, and salamander migrations are a major spring event along West Mountain Road.

DIRECTIONS:

FROM THE MASSACHUSETTS TURNPIKE: Take Exit 2 (Lee) and travel north on Route 7/20 through Lee, toward Lenox and Pittsfield. Go 6.6 miles from the turnpike exit, and turn left onto West Dugway Road, just before entering Pittsfield. Follow West Dugway Road 1.6 miles to the sanctuary entrance. ❧ FROM DOWNTOWN PITTSFIELD: Take Route 7/20 south; 1.5 miles past the Lenox town line, turn right onto West Dugway Road. Follow it 1.6 miles to the entrance.

CAPE COD AND THE ISLANDS

Ashumet Holly and Wildlife Sanctuary
286 ASHUMET ROAD
EAST FALMOUTH, MA 02536
508-563-6390

A kettlehole pond, wetlands, and old farm fields make this 49-acre sanctuary a favored spot for at least 134 species of birds, including 51 nesters. Eight different holly species and 65 varieties include American, English, Japanese, Chinese, and hybrid hollies. Rhododendrons and azaleas are scattered around the property along with other ornamentals such as dogwood and magnolia. Franklinia, a rare fall-flowering tree in the tea family discovered in Georgia in 1790, is now extinct in the wild, but survives on a few plantations such as Ashumet.

Grassy Pond is the ecological gem of Ashumet. New England coastal plain ponds are among the most unusual and threatened of Massachusetts' natural communities. Their shores contain beautiful rare plants such as the Plymouth gentian and the thread-leaved sundew in an abundance that is matched nowhere else in the world. Grassy Pond also provides critical habitat for two species of rare bluet damselflies.

DIRECTIONS

FROM ROUTE 3: Cross the Sagamore Bridge to Cape Cod and take Route 6 to the Bourne rotary. Take Route 28 south to the junction with Route 151 in North Falmouth. Follow Route 151 east for 4 miles and go left on Currier Road. The entrance to the sanctuary is 100 yards ahead on the right.

Felix Neck Wildlife Sanctuary
BOX 494, VINEYARD HAVEN, MA 02568
508-627-4850

The 350 acres of Felix Neck curve along the eastern shore of Martha's Vineyard and project into Sengekontacket Pond, which is protected from the waters of Nantucket Sound by a thin strip of sand. The beach and salt marsh associated with the pond, as well as the meadows, oak-pine forest, thickets, and cattail marshes that cover the rest of the sanctuary make for a coastal habitat that attracts over 100 species of birds. Six miles of trails wind through the neck of land that gives the sanctuary its name.

The salt marshes that border the pond around Felix Neck may be the most extensive on the island. Some small islands within the pond support rare tern colonies; those within the marshes are dominated by post oak, a species whose range barely extends into southern New England. The coastal oak woodlands on Felix Neck intergrade with the remnant sandplain grassland and coastal heathland vegetation found on nearly 40 acres of Felix Neck. This habitat supports not only some uncommon and colorful plants such as the orange butterfly-weed, but also three unusual ant species and three rare species of moths, including the large and spectacular imperial moth.

Of the 196 bird species recorded so far at Felix Neck, 62 are known to breed. Emblematic of the sanctuary are the ospreys that nest on the pole near the parking lot.

DIRECTIONS

FROM I-495: Cross the Bourne Bridge to Cape Cod and follow Route 28 south to Woods Hole. Take the ferry to Vineyard Haven. Leave Vineyard Haven on State Road, and turn left onto the Edgartown Road. The entrance to the sanctuary is on the left, 4 miles ahead.

Sampson's Island Wildlife Sanctuary
COTUIT, MA 02635
617-834-9661
(COASTAL WATERBIRD PROGRAM)

Sampson's Island is a 1.6-mile-long barrier beach featuring coastal dunes, salt marsh, and uplands of red cedar, pitch pine, and black oak on a narrow strip of land running parallel to Osterville Grand Island on Cape Cod. Terns, egrets, plovers, and night herons nest on the island, which is also heavily used by pleasure boaters in season. Massachusetts Audubon Society owns the western quarter of the island but manages the entire island as a wildlife sanctuary. From April through August, the Society's Coastal Waterbird Program monitors and protects endangered species such as least and roseate terns and piping plovers, which nest on the island during that period.

A self-guided nature trail approximately three quarters of a mile long loops through salt marsh and up-

lands before returning to the beach area. Near the water's edge are glasswort and sea lavender, interspersed with marsh grass, and farther from the water, seaside goldenrod, beach pea, beach rose, and dusty miller flourish. Bayberry and poison ivy are also common. On a narrow strip of upland, black oak, red cedar, and pitch pine dominate. Look for green herons along with flickers, towhees, mockingbirds, catbirds, and various sparrows. Along the edge of the island, bearberry, a heatherlike plant, covers parts of the dunes. Snowy egrets, great egrets, and black-crowned night herons nest in the shrubs and trees along the trail. Common and roseate terns nest in beach grass, and least terns and the piping plovers nest on the open beach.

DIRECTIONS

FROM THE MAINLAND: Access to Sampson's Island is only possible by boat from the towns of Cotuit, Osterville, and Marstons Mills. There are marinas in all three towns.

Sesachacha Heathland Wildlife Sanctuary
POLPIS ROAD, NANTUCKET, MA 02554
508-228-9208
(LOST FARM)

Nearly every acre of the nearly 900 acres on this Nantucket preserve is important to the conservation of biological diversity in Massachusetts. Under the Society's stewardship, globally threatened sandplain grassland and coastal heathland communities, the largest remaining tracts of these community types in the world, are carefully studied and managed, often with fire.

Sesachacha Pond, the largest brackish body on the island, attracts an impressive array of waterbirds in all seasons, including loons, grebes, cormorants, herons and egrets, many species of waterfowl and shorebirds, and the fish-hunting osprey. Over 300 bird species have been recorded at this sanctuary.

Periodic breaches of the barrier beach alter the water's salinity. In times of high salinity, scoters, great cormorants, and red-breasted mergansers may be present in higher numbers; with lower salinities, more canvasbacks, scaup, and gadwalls can be seen.

The salt-pond shore and barrier beach shoreline and the sanctuary's small freshwater coastal plain ponds also support an unusual flora. Among the two dozen rarities known from the sanctuary are six State Endangered plant species, including the only record for New England of the annual peanutgrass. Little bluestem grass, one of the most abundant grass species of the midwestern prairies, is dominant in some areas and mixes with big bluestem grass, switchgrass, and Indian grass. Heath species of huckleberry, bayberry, bearberry, and lowbush blueberry carpet the moors, with a sprinkling of aster and goldenrod evident in autumn.

In the coastal heathland, look for the rare black-and-white-colored barrens buckmoth flying among the scrub oaks in late October. Also be alert for short-eared owls and northern harriers slowly patrolling the heathland looking for meadow voles. The large black-tailed jackrabbit of the Great Plains has been introduced to Nantucket and can sometimes be seen. Rare plants include bushy rockrose, Nantucket shadbush, and sandplain blue-eyed grass.

DIRECTIONS

FROM ROUTE 3: Cross the Sagamore Bridge to Cape Cod and take Route 6 east to Hyannis. At Exit 6 take Route 132 to Hyannis and follow signs for Nantucket boats. Call ahead to the Steamship Authority (508-771-4000) or another carrier to confirm times and make reservations. From the rotary on Nantucket, take Polpis Road to Sesachacha Pond.

Wellfleet Bay Wildlife Sanctuary

P.O. BOX 236, OFF WEST ROAD
SOUTH WELLFLEET, MA 02663
508-349-2615

The great Wellfleet Bay estuary, stretching across the western boundaries of the wildlife sanctuary where salt marsh and tidal flats blend with the sky, ocean, and sand, forms one of the richest coastal environments in the Northeast. Its populations of birds, plants, and marine life are as varied and as changeable as the seasons.

The sanctuary's 1,200 acres include 443 acres of salt marsh, as well as acres of pinewoods, a bay beach, a freshwater pond, fields, and moor.

Over half of the salt-marsh acreage at Wellfleet Bay has never been ditched or drained. As a result, these pristine marshes and pannes are unique. Only about 5 percent of Massachusetts' 48,105 acres of salt marsh are believed to remain in such an unaltered and natural condition. At Wellfleet they now support the northernmost population of the diamond-backed terrapin. Some of the best examples of sandplain grasslands and coastal heathlands, now dwindling on the rest of the Cape, can be found at Wellfleet Bay. These natural communities, along with fairly extensive tracts of century-old pitch pine forest found on the sanctuary, also account for the impressive diversity of rare moth species (15 in all) documented for this site. Over 250 bird species have been seen at the sanctuary, and 71 of these nest. The site is a magnet for migrant songbirds and shorebirds in particular, including frequent rarities. Among the most common nesting bird species to be seen along the trails are red-winged blackbirds and eastern towhees.

The pond, marshes, and flats attract a variety of shorebirds such as black-bellied plovers, whimbrels, dowitchers, least and semipalmated sandpipers, and greater and lesser yellowlegs.

DIRECTIONS

FROM ROUTE 3: Follow Route 3 south to the Sagamore Bridge and cross onto Cape Cod. Take Route 6 east for 38 miles to North Eastham. The sanctuary entrance is on the west (bay) side of Route 6, just north of the Eastham-Wellfleet town line.

Resources

Books, articles, conservation lands, and public agencies that have special relevance to specific natural communities are listed in the main text, especially in the Places to Visit and Further Reading sections at the end of each chapter. A more general selection of resources that will help you enjoy the natural world in Massachusetts is given below. With a few exceptions the titles listed are either in print or in general circulation; many rare and useful classics of Massachusetts and New England natural history therefore are omitted.

Guides

There has never been a greater variety of relatively inexpensive natural history guides widely available in popular editions. These include illustrated guides to groups of organisms (e.g., birds, seashore life), habitats (forests, grasslands), and localities.

GUIDE SERIES

Appalachian Mountain Club Guides (trail guides, nature walks, and quiet water canoeing series). Boston: Appalachian Mountain Club.

The Audubon Society Field Guides and Nature Guides. New York: Alfred A. Knopf.

Connecticut Geological and Natural History Survey Guides. A series of bulletins, many of them containing keys and species accounts to little-known (especially invertebrate) groups; many are still available from the State of Connecticut.

Doubleday Nature Guides. Garden City, N.Y.: Doubleday & Company.

Golden Guides (actually several series including Golden Nature Guides, Golden Field Guides, and Golden Regional Guides). New York: Golden Press.

Habitat Guides. Lincoln, Mass.: Massachusetts Audubon Society.

Identification Guides. Boston: Houghton Mifflin Company.

The Peterson Field Guide Series. Boston: Houghton Mifflin Company.

Peterson First Guides. Boston: Houghton Mifflin Company.

The Pictured-Key ("How to Know...") Nature Series. Dubuque, Ia.: Wm. C. Brown Company.

Putnam's Nature Field Books. New York: G.P. Putnam's Sons.

The Sierra Club Natural Areas Guides. San Francisco: Sierra Club Books.

The Sierra Club Naturalist's Guides. San Francisco: Sierra Club Books.

Stalking the Wild Asparagus and *Stalking the Blue-eyed Scallop* (guides to wild edibles), by Euell Gibbons. New York: David McKay Company.

Stokes Nature Guides. Boston: Little, Brown and Company.

OTHER GUIDES

A Birder's Guide to Eastern Massachusetts, by Bird Observer. Colorado Springs: American Birding Association, 1994.

Birding Cape Cod, by Cape Cod Bird Club. Lincoln, Mass.: Massachusetts Audubon Society, 1990.

Birding by Ear (Eastern) (tape), by Richard K. Walton and Robert W. Lawson. Boston: Houghton Mifflin Company, 1994.

Bird Finding in New England, by Richard K. Walton. Boston: Godine, 1988.

Butterflies Through Binoculars, by Jeffrey Glassberg. Oxford: Oxford University Press, 1993.

Dragonflies and Damselflies of Cape Cod, by Ginger Carpenter. Brewster, Mass.: Cape Cod Museum of Natural History, 1991.

Field Guide to the Birds of North America. Washington, D.C.: The National Geographic Society, 1983.

Guide to Bird Sounds (tape). Ithaca, N.Y.: Cornell Laboratory of Ornithology, 1985.

A Guide to New England's Landscape, by Neil Jorgensen. Chester, Conn.: The Globe Pequot Press, 1977.

A Guide to the Properties of The Trustees of Reservations. Beverly, Mass: The Trustees of Reservations.

The Massachusetts Audubon Society Guide to Wildlife Sanctuaries, Nature Centers and Policy Offices. Lincoln, Mass.: Massachusetts Audubon Society.

Mushrooms Demystified, by David Aurora. Berkeley, Calif.: Ten Speed Press, 1986.

Newcomb's Wildflower Guide, by Lawrence Newcomb, Boston: Little, Brown and Company, 1977.

Wild Sounds of the North Woods (tape), by Lang Elliott and Ted Mack. Ithaca, N.Y.: NatureSound Studio, 1990.

Other Books

Amphibians and Reptiles of New England, by Richard M. DeGraff and Deborah D. Rudis. Amherst, Mass.: University of Massachusetts Press, 1983.

The A.O.U. Checklist of North American Birds (sixth ed.), American Ornithologists' Union, 1983; *Supplements,* including the Fortieth Supplement: *The Auk* 112(3), pp. 819^{8}830, 1995.

Aquatic Entomology, by W. Patrick McCafferty. Boston: Science Books International, 1981.

Birds of Massachusetts, by Richard R. Veit and Wayne R. Petersen. Lincoln, Mass.: Massachusetts Audubon Society, 1993.

Butterflies of Massachusetts and Other New England States, by Brian Cassie, Christopher W. Leahy, and Richard K. Walton. Lincoln, Mass.: Massachusetts Audubon Society, in press.

County Checklist of Massachusetts Vascular Plants, by Bruce A. Sorrie. Westborough, Mass.: Massachusetts Natural Heritage and Endangered Species Program, Massachusetts Division of Fisheries and Wildlife (in prep.).

Ferns and Allied Plants of New England, by A.F. Tryon and R.C. Moran. Lincoln, Mass.: Massachusetts Audubon Society, in press.

The Forests of Lilliput—The Realm of Mosses and Lichens, by John Bland. Englewood Cliffs, N.J.: Prentice-Hall, 1971.

The Freshwater Mollusks of Canada, by Arthur H. Clarke. Ottowa: National Museums of Canada, 1981.

Gray's Manual of Botany, 8th edition, revised by M.L. Fernald. New York: D. Van Nostrand Company, 1970.

Inland Fishes of Massachusetts, by K. E. Hartel, D. B. Halliwell, and A. E. Launer (in prep.).

Keys to the Freshwater Macroinvertebrates of Massachusetts (second ed.), by Douglas G. Smith. Privately printed, 1995.

Manual of the Vascular Plants of Northeastern United States and Adjacent Canada, by Henry A. Gleason and Arthur Cronquist. New York: D. Van Nostrand Company, 1963.

New England Wildlife: Habitat, Natural History, and Distribution, by Richard M. DeGraaf and Deborah D. Rudis. Tech. Rep. NE-108. Broomall, Pa.: U.S. Dept. of Agriculture, Forest Service, 1986.

Quabbin: The Accidental Wilderness, by Thomas Conuel. Revised edition. Amherst, Mass.: University of Massachusetts Press, 1990.

Rare and Endangered Vertebrates of Massachusetts, by James E. Cardoza and Thomas W. French. Lincoln, Mass.: Massachusetts Audubon Society, in press.

Roadside Geology of Vermont and New Hampshire, by Bradford B. Van Diver. Missoula, Mont.: Mountain Press, 1987. (A Massachusetts volume in the series is in preparation.)

A Synonymized Checklist of the Vascular Flora of the United States, Canada and Greenland, 2nd ed., two volumes, by John T. Kartesz. Portland, Ore.: Timber Press, 1994.

This Broken Archipelago, Cape Cod and the Islands, Amphibians and Reptiles, by James D. Lazell, Jr. New York: Quadrangle, 1976.

Wild Mammals of New England, by Alfred J. Godin. Baltimore: Johns Hopkins University Press, 1977.

Written in Stone: A Geological History of the Eastern U.S., by Chet and Maureen Raymo. Chester, Conn.: Globe Pequot Press, 1989.

Periodicals, Maps, and Other Publications

Bird Observer (field records and popular articles on Massachusetts birds). To subscribe, write to: 462 Trapelo Rd., Belmont, Mass., 02178.

"Collated List of English Names for North American Butterflies," *American Butterflies* 2 (2), pp. 28–35, 1994.

Fact Sheets on Natural Communities and Endangered Species, Massachusetts Natural Heritage and Endangered Species Program. (See under state agencies, below.)

Massachusetts Forests and Parks (map and guide to state lands). Massachusetts Department of Environmental Management, 1995.

Massachusetts Land Conservation Agencies (a listing that includes all Massachusetts land trusts), ed. by Virginia Slack. The Trustees of Reservations, 1995.

"Proposed English Names for the Odonata of North America," by Dennis R. Paulson and Sidney W. Dunkel. (Unpublished list distributed by *Argia*, the News Journal of the Dragonfly Society of the Americas, 1995.)

Remote Sensing: 20 Years of Change in Massachusetts 1951/52–1971/72, by William P. MacConnell. Research Bulletin #630, Amherst, Mass.: Massachusetts Agricultural Experiment Station, 1975.

Rhodora: Journal of the New England Botanical Club.

Sanctuary: Journal of the Massachusetts Audubon Society.

Standard Common and Current Scientific Names for North American Amphibians and Reptiles (second ed.) by J. T. Collins, R. Conant, J. E. Huheey, J. L. Knight, E. M. Rundquist and H. M. Smith. Herpetological Circular No. 12, Society for the Study of Amphibians and Reptiles, 1982.

Topographical Maps of Massachusetts, U.S. Geological Survey, Reston, Va., 22092. Also available through selected outdoor equipment and book retailers throughout the state.

Wild Flower Notes: official publication of the New England Wild Flower Society.

NOTE ON TAXONOMY, NOMENCLATURE, AND THE INDEX
In general, vernacular and scientific names follow the sources listed above: American Ornithologists' Union (1983–1995), Clarke (1981), Collins et al. (1982), Godin (1977), Hartel et al. (in prep.), Kartesz (1994), North American Butterfly Association (1994), Paulson and Dunkel (1995), Smith (1995), and Sorrie (in prep.), as well as Gosner (see p. 47) and Bigelow and Schroeder (see p. 39). The author is solely responsible for all errors and inconsistencies.

This book is intended for a general readership, and therefore we have—with very few exceptions—used only vernacular names in the main text. However, because botanists and invertebrate specialists are often more conversant with scientific names than with common ones, we have indexed all plants and invertebrates mentioned in the text in a vernacular/scientific format.

Conservation Organizations, Programs, and Clubs

Appalachian Mountain Club
5 Joy Street
Boston, MA 02108

Bird Clubs: There are at least 15 active bird clubs in Massachusetts, but no directory. For details, contact the Massachusetts Audubon Society.

Boston Mycological Club
Holliston, MA 01746

Center for Biological Conservation
Massachusetts Audubon Society
208 South Great Road
Lincoln, MA 01773

Center for Coastal Studies
59 Commercial Street
Provincetown, MA 02657

The Environmental Institute
Blaisdell House, University of Mass.
Amherst, MA 01003

Environmental League of Mass. (ELM)
3 Joy Street
Boston, MA 02108

Hitchcock Center for the Environment
525 South Pleasant Street
Amherst, MA 01002

Lloyd Center for Environmental Studies
430 Potomska Road
South Dartmouth, MA 02748

Manomet Observatory for Conservation
Sciences, Box 936
Manomet, MA 02345

Maria Mitchell Science Center
2 Vestal Street
Nantucket, MA 02554

Massachusetts Association of Conservation Commissions (MACC)
10 Juniper Road
Belmont, MA 02178

Massachusetts Audubon Society
208 South Great Road
Lincoln, MA 01773

Massachusetts Forestry Association
Box 1096, Belchertown, MA 01007

Massachusetts Land Trust Coalition (presently chaired by The Trustees of Reservations). There are about 160 land trusts in the Commonwealth, some of which own thousands of acres locally and publish guides to their holdings. For a listing, see *Massachusetts Land Conservation Agencies* under Periodicals, above.

The Nature Conservancy
79 Milk Street
Boston, MA 02109

New England Botanical Club
22 Divinity Avenue
Cambridge, MA 02138

New England Wild Flower Society
Hemenway Road
Framingham, MA 01701

Nuttall Ornithological Club
Harvard University
Cambridge, MA 02138

QLF/Atlantic Center for the Environment
39 South Main Street
Ipswich, MA 01938

South Shore Natural Science Center
Jacob's Lane
Norwell MA 02061-0429

The Trustees of Reservations
572 Essex Street
Beverly, MA 01915

Museums and Living Collections

The Berkshire Museum
39 South Street
Pittsfield, MA 01201

Cape Cod Museum of Natural History
Route 6A
Brewster, MA 02631

The Children's Museum (Boston)
300 Congress Street
Boston, MA 02210

The Fisher Museum at Harvard Forest
Route 32
Petersham, MA 01366

The Garden in the Woods
Hemenway Road
Framingham, MA 01701

The Gray Herbarium
Harvard University
22 Divinity Street
Cambridge, MA 02138

The Harvard Museum of Cultural and Natural History (includes the Museum of Comparative Zoology)
Harvard University
26 Oxford Street
Cambridge, MA 02138

Museum of Science
Science Park
Boston, MA 02114

New England Aquarium
Central Wharf
Boston, MA 02110

New England Science Center
222 Harrington Way
Worcester, MA 01604

Peabody Essex Museum
East India Square
Salem, MA 01970

Pratt Museum of Natural History
Amherst College
Amherst, MA 01002

Springfield Science Museum
236 State Street
Springfield, MA 01085

Government Environmental Agencies and Programs

Environmental Protection Agency (EPA)
1 Congress Street
Boston, MA 02203

Executive Office of Environmental Affairs (EOEA)
100 Cambridge Street
Boston, MA 02202

Massachusetts Coastal Zone Management Office
100 Cambridge Street
Boston, MA 02202

Massachusetts Coop. Extension Service
University of Massachusetts
Amherst, MA 01003

Massachusetts Department of Environmental Management (DEM) (including Division of Forests & Parks)
100 Cambridge Street
Boston, MA 02202

Massachusetts Department of Environmental Protection (DEP)
One Winter Street
Boston, MA 02108

Massachusetts Division of Fisheries and Wildlife (Field Headquarters)
1 Rabbit Hill Road
Westborough, MA 01581

Massachusetts Div. of Marine Fisheries
100 Cambridge Street
Boston, MA 02202

Massachusetts Natural Heritage and Endangered Species Program
1 Rabbit Hill Road
Westborough, MA 01581

Massachusetts Water Resources Authority
Charlestown Navy Yard
100 First Street
Boston, MA 02129

Metropolitan District Commission
20 Somerset Street
Boston, MA 02108

National Park Service, N. Atlantic Region
15 State Street
Boston, MA 02109

U.S. Fish and Wildlife Service
Northeast Regional Office
300 Westgate Center Drive
Hadley, MA 01035

Glossary

This short glossary is not intended to be a comprehensive ecological dictionary, but simply a key to technical terms and jargon that appear occasionally in the text despite our best efforts to banish them. Terms defined in the text itself and indexed are not included here.

ACEC. Area of Critical Environmental Concern. An area of open space that has been formally defined by the Commonwealth of Massachusetts as of special value for its natural characteristics. Candidate localities proposed by individuals or municipalities must meet certain criteria and be endorsed by both biologists and members of the community.

ACHENE. A dry one-seeded fruit formed from a single female flower (e.g., maple samara). The achenes of sedges are sometimes useful in identification.

ANADROMOUS. Living in the sea but returning to fresh water to breed (e.g., Atlantic salmon). Compare *Catadromous.*

CATADROMOUS. Living in fresh water but returning to the sea to breed (e.g., American eel). Compare *Anadromous.*

COMMUNITY. As an ecological term, an assemblage of organisms found together often enough to give them a distinct collective identity. As a biological concept, the natural community is less distinct, since many organisms are not confined within the boundaries of a single community and the boundaries between communities are often blurred. The concept is nonetheless useful as a conservation and interpretive tool because it allows us to define the natural world (however crudely) in terms of the landscape, rather than as habitat for individual organisms. The Nature Conservancy and state natural heritage programs have developed and used the natural community idea with great success. To the reader of this book it will become clear that some communities (e.g., bogs) are easier to define sharply than others (e.g., various forest types) because they have more obligate species and share fewer species with other communities. It will also be obvious that all communities are not created equal: the plant community of a coastal plain pond shore, though of great interest and value, exists at a different scale than the ocean. This simply reflects the inherent complexity of the natural world and the concomitant challenge of describing it. Compare *Ecosystem, Habitat,* and *Niche.*

COSMOPOLITAN. Of worldwide distribution. For example, peregrine falcon and Norway rat occur on all of the continents except Antarctica. Compare *Endemic.*

DOMINANT. Used here to describe the most abundant species of certain plant communities.

ECOSYSTEM. A functioning unit of the environment in which both living and nonliving components interact. The term may be applied at any scale, e.g., the global ecosystem, the rain forest ecosystem. See also *Community, Habitat,* and *Niche.*

ENDEMIC. Restricted in distribution to a particular area. For example, the bird family containing all the wood warblers (Parulidae) is endemic to the western hemisphere, while the Muskegat vole is endemic to Muskegat Island near Nantucket.

ERICACEOUS. Belonging to the heath family (Ericaceae), e.g., blueberries, cranberries, heathers, laurels, and rhododendrons. Also referring broadly to low, shrubby aromatic vegetation such as that of heathlands or moorlands.

EXOTIC. A species that has been imported by human agency—either intentionally or accidentally—into an environment other than the one in which it naturally occurs. *Alien* and *nonnative* are frequently used synonyms. Some of these living imports (e.g., ox-eye daisy, brown trout) seem relatively benign, apparently filling an empty ecological niche without displacing any native species. Many other species, such as Norway rats, European starlings, zebra mussels, Gypsy moths, and purple loosestrife, compete with native species, destroy habitat, or cause problems for people.

EXTIRPATED. Deliberately exterminated. The passenger pigeon and many other species were wiped out by human beings. This form of extinction is now occurring at a rate at least a thousand times more rapid than the major extinctions documented in the fossil record.

EXUVIAE or **EXUVIUM.** The cast-off skin of an animal. Commonly used for the nymphal exoskeleton of insects such as dragonflies once the adult form has emerged, but also correct in referring to a sloughed snake skin.

FACULTATIVE SPECIES. A species that frequently occurs in a particular community under certain conditions but can also survive in other habitats. Compare *Obligate* and *Indicator species.*

FEDERALLY ENDANGERED/THREATENED. These are legally pertinent concepts defined as part of the Federal Endangered Species Act:

> *Endangered:* Any species that is in danger of extinction throughout all or a significant portion of its range.

> *Threatened:* Any species that is likely to become an endangered species within the foreseeable future throughout all or a significant portion of its range.

FOLIOSE. Leafy, leaf-like. Referring especially to certain lichens (e.g., shield lichen and rock tripes).

FORBS. Collective term for herbaceous flowering plants that are not grasses, sedges, or rushes; wildflowers.

FRUTICOSE. Shrublike, bushy. Fruticose lichens are upright and branched species such as British soldiers.

HABITAT. The place in which a given organism prefers to live. This may be quite broad (e.g., Arctic tundra) or very restricted (e.g., the canopy of oak trees). The emphasis is on behavior of the organism. Compare *Community, Ecosystem,* and *Niche.*

HERB, HERBACEOUS. As used here, refers to plants that lack a persistent woody stem above the ground, i.e., plants other than trees, shrubs, and woody vines. Not restricted here to the narrower definition of plants used as food, seasoning, or medicine.

HYDRIC. Wet; characterized by high moisture content. Refers especially to the soil conditions that typify marshes and other wetlands. Compare *Mesic* and *Xeric.*

INDICATOR SPECIES. Species that are characteristic of a given natural community/habitat. The strongest indicators are those most closely associated with their community. If you are looking at a piping plover nest in Massachusetts, you must be standing on a barrier beach. However, other species may be indicators of more than one habitat and be present in others without necessarily being typical. See also *Obligate* and *Facultative species.*

IRRUPTION. An irregular movement by numbers of animals (especially birds) following the breeding season into areas beyond their normal range due to an inadequate food supply (e.g., the appearance in late fall and winter in Massachusetts of owls, finches, and waxwings from the North and West).

KETTLEHOLE. A typically pot-shaped depression in the earth formed by the melting of a chunk of glacial ice that had been buried in sediments. When the depression extends below the water table it becomes a kettlehole pond. Kettleholes provide the structure for a number of natural communities such as bogs.

LENTIC. Still, referring to waters such as ponds and lakes. Compare *Lotic.*

LOTIC. Moving, referring to waters such as rivers and streams. Compare *Lentic.*

MACROINVERTEBRATES. Larger forms of invertebrate life such as insects, as distinguished from microscopic animals such as amoebas.

MESIC. Moist or requiring a moderate amount of moisture. Refers to soil conditions, habitats (e.g., rich mesic forest), and growing conditions favored by species of plants. Compare *Hydric* and *Xeric.*

NICHE. The ecological role a species plays within its environment, not just where it lives, but what it eats and how it interacts with the other constituents of the ecosystem, both living and nonliving. Compare *Community, Ecosystem,* and *Habitat.*

OBLIGATE SPECIES. Species restricted to or dependent upon a particular habitat or natural community (e.g., the sperm whale in the deep ocean or marsh wrens in freshwater marshes). Compare *Indicator species* and *Facultative species.*

OLIGOTROPHIC. Low in nutrients. Describes deep, cold lakes or highly acid bodies such as bogs where plant growth is inhibited.

ORGANIC SOIL. Soil composed of at least 12 percent organic carbon, in some cases up to 98 percent (that is, made up to a significant degree of decomposed plant tissue); the remaining components may be clay, sand, or other minerals. Dried organic soils will burn.

PELAGIC. Referring to deep waters (>20 meters) of a lake or the sea and the organisms (e.g., shearwaters, striped bass) that inhabit such waters, as distinct from those that live on the bottom (benthic).

PERIGYNIUM. A sac that encloses the female flower in the sedge genus *Carex.* The characteristics of the perigynia are crucial in the identification of this notoriously difficult group.

RIPARIAN. Referring to the banks of a river or other water body.

RUSH. Member of a relatively small plant family, the Juncaceae. Often confused with grasses and sedges because of general similarity (linear leaves, inconspicuous flowers) and because of colloquial names such as bulrush. The fruits are distinctively shaped, three-sided capsules. See also *Sedge.*

SEDGE. Member of a large plant family, the Cyperaceae. Distinguished from grasses, with which they are often confused, by their solid (not hollow) jointless stems, which are often, but not always, triangular (remember: sedges have edges). An ecologically important group of plants with many species endemic to

specific (especially wetland) habitats. Notoriously difficult to identify. See also *Rush.*

STATE ENDANGERED/THREATENED/SPECIAL CONCERN. Legally pertinent concepts defined as part of the Massachusetts Endangered Species Act, M.G.L. c. 131A:

Endangered: Any species of plant or animal in danger of extinction throughout all or a significant portion of its range, and species of plants or animals in danger of extirpation as documented by biological research and inventory.

Threatened: Any species of plant or animal likely to become an endangered species within the foreseeable future throughout all or a significant portion of its range, and any species declining or rare as determined by biological research and inventory and likely to become endangered in the foreseeable future.

Special Concern: Any species of plant or animal that has been documented by biological research and inventory to have suffered a decline that could threaten the species if allowed to continue unchecked or that occurs in such small numbers or with such a restricted distribution or specialized habitat requirements that it could easily become threatened within Massachusetts.

Copies of the Massachusetts Endangered Species list and accompanying regulations can be obtained from the State House Bookstore, Room 116, Boston, MA 02133. See also *Federally Endangered/Threatened.*

SWALE. A low-lying wet area, sometimes seasonally flooded. Often dominated by grasses, sedges, and wetland shrubs such as cranberry.

SWAMP. A wooded wetland, capable of supporting trees or shrubs (e.g., buttonbush swamp or red maple swamp).

TRUE BUGS. In the strictest sense the term refers to members of the insect order Hemiptera including water bugs, water striders, stink bugs, assassin bugs, bed bugs, and many others. Sometimes extended to include the Homoptera, e.g., planthoppers, cicadas, and aphids.

UMBELLIFER. Any member of the carrot family (Umbelliferae), named for the umbel form of the flower cluster in which all of the flower stalks radiate from the same point like the spokes of an umbrella.

UPLANDS. Elevated, well-drained land as distinguished from wetlands. Not to be confused with highlands, which is land of high elevation.

WETLANDS. An area of wet soils, usually flooded or saturated for at least part of the year. Different types of wetlands are typified by plant species capable of tolerating different degrees of flooding and water chemistry.

WRACK, WRACK LINE. Species of seaweed (e.g., knotted wrack), but also a collective term for accumulations of dead seaweed on the shore. The wrack line that typically marks the highest tide line on a barrier beach is a thriving microcommunity that supports many species of flies, isopods, and other invertebrates that feed on and breed in the rotting algae. These in turn are an important source of food for shorebirds and other predaceous species.

XERIC. Dry. Refers especially to soils and associated plant species adapted to survive with little moisture. It is a relative term that applies to deserts, but also, for example to moderately dry plant communities on exposed, south-facing slopes. Compare *Hydric* and *Mesic.*

Index

pumpkinseed (fish), 130
purslane, water *(Ludwigia palustris)*, 129
pussytoes *(Antennaria neglecta)*, 155
pygmyweed, water *(Tillaea aquatica)*, 51, 55
pyrola: green-flowered pyrola *(Pyrola virens)*,
   99; one-sided pyrola *(Orthilia secunda)*,
   99; pink pyrola *(P. asarifolia* var. *purpurea)*,
   183; round-leaved pyrola *(P. americana)*,
   99; shinleaf *(P. elliptica)*, 99

Quabbin Reservoir, 134
quackgrass *(Agropyron)*: quackgrass
   (A. repens), 155; seabeach quackgrass
   (A. litorale), 54
quartz, 49
Queen Anne's lace *(Daucus carota)*, 155
quill, black (early brown spinner)
   (Leptophlebia cupida), 107, 141
quill Gordon *(Epeorus pleuralis)*, 141
quillworts *(Isoetes* spp.), 129

raccoon, 100, 102, 107, 132, 148, 194, 197,
   199, 201
racer, northern black, 100, 157
radish, wild *(Raphanus raphanistrum)*, 54,
   155
ragged robin *(Lychnis flos-cuculi)*, 155
ragweed *(Ambrosia)*: common
   (A. artemisiifolia), 156; great *(A. trifida)*,
   148
rail, 196; clapper, 65; king, 121; Virginia,
   121, 191
raised bogs, 176
raisin, wild *(Viburnum cassinoides)*, 54
ramshorn, large eastern *(Helistoma t.
   trivolis)*, 129
raptors, 57, 64, 79, 87, 148, 201. *See also*
   hawks
raspberry, wild *(Rubus idaeus)*, 54
rattle, yellow *(Rhinanthus minor)*, 155
rattlesnake-plantain *(Goodyera)*, 202; downy
   (G. pubescens), 99
razorbill, 37
red admiral *(Vanessa atalanta)*, 36, 55, 156
red maple samaras, **111**
red maple swamps, 25, 110-15, 194, 199, 200
red tides, 35
redfish (rosefish), 37
redhead (duck), 130
red-spotted purple *(Limenitis arthemis
   astyanax)*, 100
redstart, American, 100
redtop *(Agrostis alba)*, 155
redweed, tufted *(Gigartina stellata)*, 44
reed, common *(Phragmites australis)*, 61, 63,
   66, 119, 122, 184, 198, 201
reindeer moss (lichen) *(Cladonia
   rangiferina)*, 55, 188
reptiles: of barrier beach and dunes, 56; of
   bogs, 177; of calcareous fens, 183; of
   coastal heathlands, 87; of cultural
   grasslands, 157; of freshwater marshes,
   119; of lakes and ponds, 130; of
   northern hardwood forests, 168; of oak-
   conifer forests, 100; of the ocean, 37;
   of pitch pine-scrub oak barrens, 93; of
   red maple swamps, 114; of rivers and
   streams, 142; of salt marshes, 64; of
   vernal pools, 107
rheotaxis, 138
Rhode Island, 71
rhododendrons, 203

rhodora *(Rhododendron canadense)*, 176
ribbed pod *(Siliqua costata)*, 55
rice: cultivated *(Oryza sativa)*, 152; wild
   (Zizania aquatica), 119, 140
rice grass, mountain *(Oryzopsis racemosa)*,
   172
rich mesic forests, 170-73, 203
Richards, William Trost, 44
ringlet, inornate (common) *(Coenonympha
   tullia)*, 156
Rio Bravo Conservation and Management
   Area (Belize), 20
rivers and streams, 137-45
robberfly, great reddish *(Proctocanthus
   rufus)*, 56
robin, American, 16, 100, 157, 201
rock tripe (lichen), 196
rock-cress (Drummond's arabis) *(Arabis
   drummondii)*, 54
rockets: dame's *(Hesperis matronalis)*, 155;
   sea *(Cakile edentula)*, 54
Rockport, Mass., 47
rockrose, bushy *(Helianthemum dumosum)*,
   79, 86, 205
rockweed *(Ascophyllum* spp. and Fucus spp.),
   35, 43, 47
rockweed (brown seaweed) zone, 43
rocky shore and intertidal zone, 41-47, 50
Roosevelt, Teddy, 7
rose *(Rosa)*: multiflora *(R. multiflora)*, 154;
   salt spray (beach) *(R. rugosa)*, 54, 193, 204
rosemary, bog *(Andromeda polifolia* var.
   glaucophylla), 176, 183, 194
rotifers (wheel animalcules) (Phylum
   Rotatoria), 127
rubyspot, American *(Hetaerina americana)*,
   141
rudd, 130, 142
rue, goat's *(Tephrosia virginiana)*, 79
ruff, 194
Rumney Marsh, 65
running pine (clubmoss) *(Lycopodium
   complanatum)*, 99
rush *(Juncus)*: brackish rush *(J. balticus
   var. littoralis)*, 54, 63; Canada (marsh)
   rush *(J. canadensis)*, 54, 119, 129;
   Greene's rush *(J. greenei)*, 54; jointed
   rush *(J. articulatus)*, 54; mud-fruited
   rush *(J. pelocarpus)*, 107; saltmarsh rush
   (blackgrass) *(J. gerardii)*, 63; seaside
   toad-rush *(J. bufonius* var. *halophilus)*, 63;
   sharp-fruited rush *(J. acuminatus)*, 54;
   short-tailed rush *(J. brevicaudatus)*, 119;
   soft rush *(J. effusus)*, 54, 119, 129; toad
   rush *(J. bufonius* var. *bufonius)*, 54. *See
   also* wood rush
rye, riverbank (wild) *(Elymus riparius)*,
   140, 148

sage, wood (American germander) *(Teucrium
   canadense)*, 55, 148
St. John's-wort *(Hypericum)*, 107; common
   (H. perforatum), 155; marsh *(H. virginicum)*,
   119; northern *(H. boreale)*, 55
St. Lawrence River, 33
salamanders, 106, 203; blue-spotted, 107,
   191; four-toed, 107, 114, 177;
   Jefferson's, 107, 114; marbled, 106,
   106, 107, 108, 114; mole (Family
   Ambystomatidae), 100, 106, 108, 148,
   166, 191, 192; northern dusky, 100, 142,
   144, 145; northern two-lined, 114, **144;**

redback, 100, **103,** 114; spotted, 106,
   107, **109,** 114, 191, 194, 199; spring,
   114, **139,** 142; two-lined, 142
salinity, of ocean, 31
sallow, blueberry *(Apharetra purpurea)*, 93
salmon, Atlantic, 37, 142
salt marshes, 16, 49, 61-67, 150, 204, 205
salt meadows, 150
salt ponds, 51
saltmarsh mosquito *(Aedes sollicitans)*, 64
saltwort: common *(Salsola kali)*, 54; dwarf,
   64
samphire (common glasswort) *(Salicornia
   europaea)*, 63, 204
Sampson's Island Wildlife Sanctuary, 204
sand dollar *(Echinarachnius parma)*, 37
sand dunes: formation of, 49; waves on
   beaches, 30. *See also* barrier beach and
   dunes
sand lance, 37
sand sedge *(Bulbostylis capillaris)*, 54
sanddragon, common *(Progomphus obscurus)*,
   130, 141
sanderling, **48,** 52, 56
sandpiper, 46, 51, 52; Baird's, 157, 194;
   buff-breasted, 157, 194; least, 57, 64,
   194, 205; purple, **40,** 46; semipalmated,
   56, 205; solitary, 107; spotted, 132,
   138, 142; upland, **76-77,** 79, 153, 157,
   158; western, 56-57; white-rumped, 57
sandplain grasslands, 76-81, 152, 204, 205
sand-shell (mussel), pointed *(Ligumia
   nasuta)*, 70, 129, 140
sand-spurrey, saltmarsh *(Spergularia
   marina)*, 63
sandworts: grove *(Moehringia lateriflora)*,
   54; seabeach *(Arenaria peploides)*, 55
Sandy Neck, 57, 58, 67
sapsucker, yellow-bellied, 100, 168, 192,
   202, 203
sarsaparilla *(Aralia)*: bristly *(A. hispida)*, 55;
   wild *(A. nudicaulis)*, 55, 99, 188
sassafras *(Sassafras albidum)*, 51, 54, 192, 193
Saugus, Mass., 65
scabious, field *(Knautia arvensis)*, 156
scallops: bay *(Aequipecten irradians)*, 35;
   deep-sea *(Placopecten magellanicus)*, 36
scaup, 204; greater, 130; lesser, 130
Schroeder, William C., 37
Scituate, Mass., 47, 65
scoter, 204; black, 46, 195; surf, 46, 195;
   white-winged, 46, 195
scuds (grammarid amphipods) (esp. *Gam-
   marus oceanicus* and *G. mucronatus)*, 44
sculpin: longhorn, 37; shorthorn, 37; slimy,
   130, 142
sea anemone. *See* anemones
sea butterfly: naked *(Clione limacina)*, 36;
   shelled (plankter) *(Limacina
   retroversa)*, 36
sea cucumbers (Phylum Echinodermata,
   Class Holothuroidea), 36, 44; silky (sea)
   cucumber *(Chiridota laevis)*, 36
sea ducks, 196. *See also* bufflehead; ducks,
   harlequin; eider; goldeneye; merganser;
   scoter
sea gooseberry *(Pleurobrachia pileus)*, 36
sea grapes *(Molgula* spp.), 37, 46
sea lace (lacy crusts) (Phylum Bryozoa, esp.
   *Calopora craticula, Electra pilosa,
   Membranipora* spp., and *Tegella unicornis*
   spp.), 36, 44